Photovoltaic Systems

Photovoltaic Systems

Jaime Hunter

Larsen & Keller
www.larsen-keller.com

Photovoltaic Systems
Jaime Hunter
ISBN: 978-1-64172-654-2 (Hardback)

Larsen & Keller
Published by Larsen and Keller Education,
5 Penn Plaza,
19th Floor,
New York, NY 10001, USA

Cataloging-in-Publication Data

Photovoltaic systems / Jaime Hunter.
p. cm.
Includes bibliographical references and index.
ISBN 978-1-64172-654-2
1. Photovoltaic power systems. 2. Power-plants.
3. Photovoltaic power generation. I. Hunter, Jaime.
TK1087 .P46 2022
621.312 44--dc23

For more information regarding Larsen and Keller Education and its products, please visit the publisher's website www.larsen-keller.com

TABLE OF CONTENTS

PREFACE

This book aims to help a broader range of students by exploring a wide variety of significant topics related to this discipline. It will help students in achieving a higher level of understanding of the subject and excel in their respective fields. This book would not have been possible without the unwavered support of my senior professors who took out the time to provide me feedback and help me with the process. I would also like to thank my family for their patience and support.

The conversion of sunlight into electricity is known as photovoltaic. A power system, which is designed to supply the usable solar power through photovoltaics is referred to as a photovoltaic system. It is made up of an arrangement of many components such as solar panels and solar inverter. Solar panels are used for the absorption and conversion of sunlight into electricity. Solar inverters convert the output from direct to alternating current. Some of the other components of photovoltaic systems are solar array, mounting, cabling, tracker and batteries. Photovoltaic systems can range from small PV systems such as rooftop-mounted or building-integrated systems to large utility-scale power stations. The topics included in this book on photovoltaic systems are of utmost significance and bound to provide incredible insights to readers. Most of the topics introduced herein cover new techniques and applications of photovoltaic systems. This textbook will provide comprehensive knowledge to the readers.

A brief overview of the book contents is provided below:

Chapter - Introduction

Photovoltaics refers to the conversion of light into electricity with the help of semiconducting materials which exhibit the photovoltaic effect. The creation of electric current and voltage in a material upon exposure to light takes place due to photovoltaic effect. This is an introductory chapter which will briefly introduce all the significant aspects of photovoltaics.

Chapter - Photovoltaics: Concepts and Techniques

Some of the common concepts of photovoltaics are solar radiation, solar irradiance, solar constant, solar cell efficiency, etc. A number of techniques are used in photovoltaics such as sun-free photovoltaics, polarizing organic photovoltaics and cadmium telluride photovoltaics. This chapter has been carefully written to provide an easy understanding of these concepts and techniques of photovoltaics.

Chapter - Photovoltaic Systems

Photovoltaic systems are the power systems that are designed to supply solar power by means of photovoltaics. The components of photovoltaic systems include solar cell, solar panel, solar tracker and solar inverter. The topics elaborated in this chapter will help in gaining a better perspective about photovoltaic systems and its components.

Chapter - Electricity Generation Systems

The process that generates electric power from primary energy sources is referred to as electricity generation. Some of the electricity generation systems are photovoltaic power stations, grid-connected photovoltaic power systems, rooftop photovoltaic power station and building-integrated photovoltaics.

This chapter closely examines these types of electricity generation systems to provide an extensive understanding of the subject.

Chapter - Applications

Photovoltaics are widely used in solar-powered refrigerators, solar lamps, solar-powered fans, solar vehicles, solar panels on spacecrafts, solar-powered pumps, solar-powered watches, solar-powered flashlights and solar air conditioners. All these diverse applications of photovoltaics have been carefully analyzed in this chapter.

Jaime Hunter

Introduction

CHAPTER 1

Photovoltaics refers to the conversion of light into electricity with the help of semiconducting materials which exhibit the photovoltaic effect. The creation of electric current and voltage in a material upon exposure to light takes place due to photovoltaic effect. This is an introductory chapter which will briefly introduce all the significant aspects of photovoltaics.

SOLAR ENERGY

Solar energy is radiation from the Sun capable of producing heat, causing chemical reactions, or generating electricity. The total amount of solar energy incident on Earth is vastly in excess of the world's current and anticipated energy requirements. If suitably harnessed, this highly diffused source has the potential to satisfy all future energy needs. In the 21st century solar energy is expected to become increasingly attractive as a renewable energy source because of its inexhaustible supply and its non-polluting character, in stark contrast to the finite fossil fuels coal, petroleum, and natural gas.

The Sun is an extremely powerful energy source, and sunlight is by far the largest source of energy received by Earth, but its intensity at Earth's surface is actually quite low. This is essentially because of the enormous radial spreading of radiation from the distant Sun. A relatively minor additional loss is due to Earth's atmosphere and clouds, which absorb or scatter as much as 54 percent of the incoming sunlight. The sunlight that reaches the ground consists of nearly 50 percent visible light, 45 percent infrared radiation, and smaller amounts of ultraviolet and other forms of electromagnetic radiation.

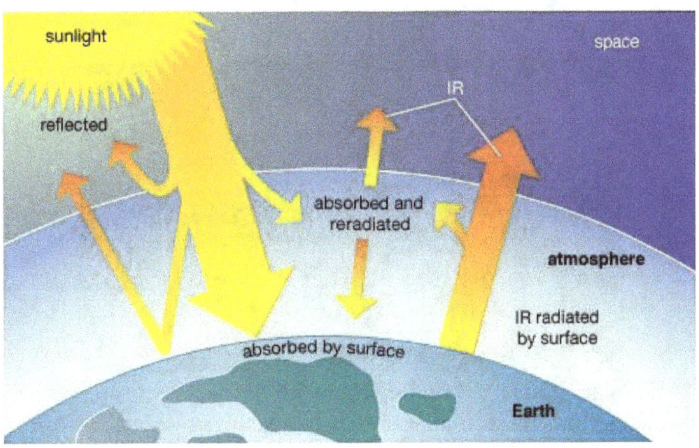

Solar energy: Reflection and absorption of solar energy. Although some incoming sunlight is reflected by Earth's atmosphere and surface, most is absorbed by the surface, which is warmed.

Uses of Solar Energy

The potential for solar energy is enormous, since about 200,000 times the world's total daily electric-generating capacity is received by Earth every day in the form of solar energy. Unfortunately, though solar energy itself is free, the high cost of its collection, conversion, and storage still limits its exploitation in many places. Solar radiation can be converted either into thermal energy (heat) or into electrical energy, though the former is easier to accomplish.

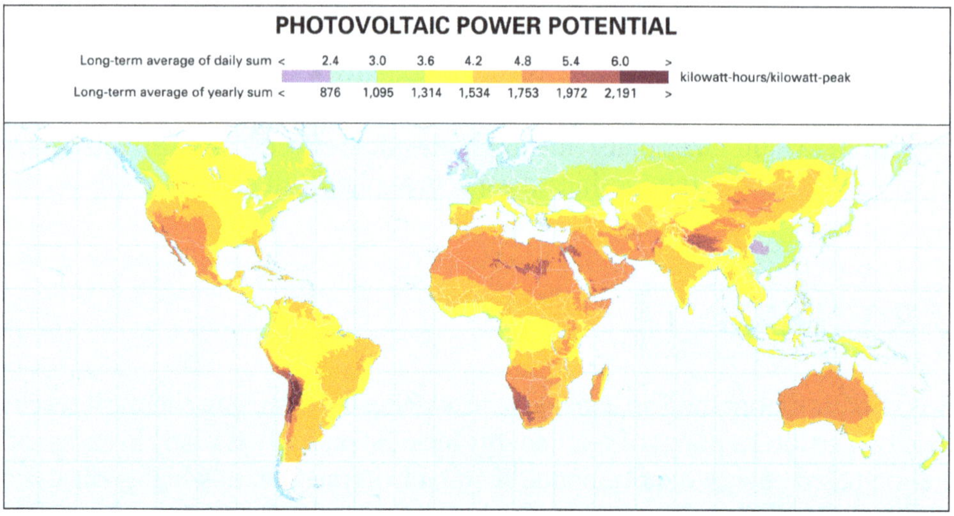

Solar energy potential, Earth's photovoltaic power potential.

Thermal Energy

Among the most common devices used to capture solar energy and convert it to thermal energy are flat-plate collectors, which are used for solar heating applications. Because the intensity of solar radiation at Earth's surface is so low, these collectors must be large in area. Even in sunny parts of the world's temperate regions, for instance, a collector must have a surface area of about 40 square metres (430 square feet) to gather enough energy to serve the energy needs of one person.

Solar heating: A building roof with flat-plate collectors that capture solar energy to heat air or water.

The most widely used flat-plate collectors consist of a blackened metal plate, covered with one or two sheets of glass, that is heated by the sunlight falling on it. This heat is then transferred to air or water, called carrier fluids, that flow past the back of the plate. The heat may be used directly, or it may be transferred to another medium for storage. Flat-plate collectors are commonly used for solar water heaters and house heating. The storage of heat for use at night or on cloudy days is commonly accomplished by using insulated tanks to store the water heated during sunny periods. Such a system can supply a home with hot water drawn from the storage tank, or, with the warmed water flowing through tubes in floors and ceilings, it can provide space heating. Flat-plate collectors typically heat carrier fluids to temperatures ranging from 66 to 93 °C (150 to 200 °F). The efficiency of such collectors (i.e., the proportion of the energy received that they convert into usable energy) ranges from 20 to 80 percent, depending on the design of the collector.

Another method of thermal energy conversion is found in solar ponds, which are bodies of salt water designed to collect and store solar energy. The heat extracted from such ponds enables the production of chemicals, food, textiles, and other industrial products and can also be used to warm greenhouses, swimming pools, and livestock buildings. Solar ponds are sometimes used to produce electricity through the use of the organic Rankine cycle engine, a relatively efficient and economical means of solar energy conversion, which is especially useful in remote locations. Solar ponds are fairly expensive to install and maintain and are generally limited to warm rural areas.

On a smaller scale, the Sun's energy can also be harnessed to cook food in specially designed solar ovens. Solar ovens typically concentrate sunlight from over a wide area to a central point, where a black-surfaced vessel converts the sunlight into heat. The ovens are typically portable and require no other fuel inputs.

A monk using a solar-powered cookstove in the Potala Palace, Lhasa, Tibet.

Electricity Generation

Solar radiation may be converted directly into electricity by solar cells (photovoltaic cells). In such cells, a small electric voltage is generated when light strikes the junction between a metal and a semiconductor (such as silicon) or the junction between two different semiconductors. The power generated by a single photovoltaic cell is typically only about two watts. By connecting large numbers of individual cells together, however, as in solar-panel arrays, hundreds or even thousands of kilowatts of electric power can be generated in a solar electric plant or in a large household array. The energy efficiency of most present-day photovoltaic cells is only about 15 to 20 percent, and, since the intensity of solar radiation is low to begin with, large and costly assemblies of such cells are required to produce even moderate amounts of power.

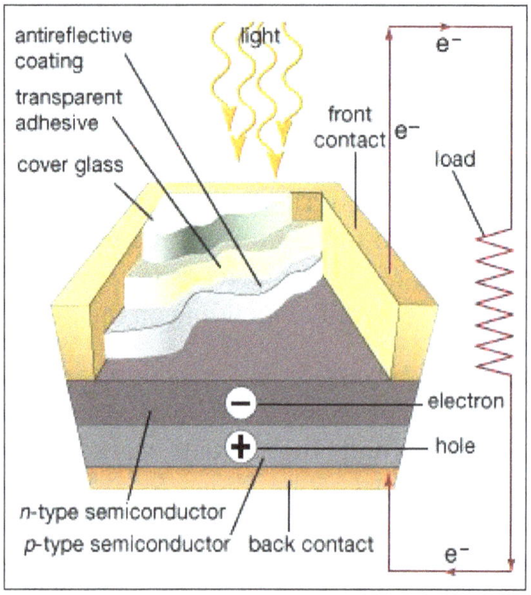

When sunlight strikes a solar cell, an electron is freed by the photoelectric effect. The two dissimilar semiconductors possess a natural difference in electric potential (voltage), which causes the electrons to flow through the external circuit, supplying power to the load. The flow of electricity results from the characteristics of the semiconductors and is powered entirely by light striking the cell.

Small photovoltaic cells that operate on sunlight or artificial light have found major use in low-power applications—as power sources for calculators and watches, for example. Larger units have been used to provide power for water pumps and communications systems in remote areas and for weather and communications satellites. Classic crystalline silicon panels and emerging technologies using thin-film solar cells, including building-integrated photovoltaics, can be installed by homeowners and businesses on their rooftops to replace or augment the conventional electric supply.

Solar power: Single-family house with solar panels on the roof.

Concentrated solar power plants employ concentrating, or focusing, collectors to concentrate sunlight received from a wide area onto a small blackened receiver, thereby considerably increasing the light's intensity in order to produce high temperatures. The arrays of carefully aligned mirrors or lenses can focus enough sunlight to heat a target to temperatures of 2,000 °C (3,600 °F) or more. This heat can then be used to operate a boiler, which in turn generates steam for a steam turbine electric generator power plant. For producing steam directly, the movable mirrors can be arranged so as to concentrate large amounts of solar radiation upon blackened pipes through which water is circulated and thereby heated.

Concentrated solar-power plant: Nevada Solar One, a concentrated solar-power plant.

Solar reflector: One of the reflectors at the Solar Two power plant, Mojave Desert, Calif.

Other Applications

Solar energy is also used on a small scale for purposes other than those described above. In some countries, for instance, solar energy is used to produce salt from seawater by evaporation. Similarly, solar-powered desalination units transform salt water into drinking water by converting the Sun's energy to heat, directly or indirectly, to drive the desalination process.

Solar technology has also emerged for the clean and renewable production of hydrogen as an alternative energy source. Mimicking the process of photosynthesis, artificial leaves are silicon-based devices that use solar energy to split water into hydrogen and oxygen, leaving virtually no pollutants. Further work is needed to improve the efficiency and cost-effectiveness of these devices for industrial use.

PHOTOVOLTAICS

Photovoltaic (PV) essentially means electricity from the energy of the sun and is derived from the words "photo" with the meaning light and "voltaic" meaning voltage. The term "photovoltaic" is used to describe a process known as the "photovoltaic effect" the process by which a material such as silicon converts sunlight into electricity.

PV cells are laminated to a thin sheet of glass. Sunlight in the form of photons strike the PV cells, exciting electrons in the silicon. An electric current is generated when the electrons are collected by wiring that is incorporated into the PV cells.

A solar panel consists of multiple PV cells, the amount will depend on the output voltage and current required of the solar panel. Solar panels range in size from around 5 to over 400watts peak output and produce DC (direct current).

Multiple solar panels are connected in strings, to build a solar array. The size of the array will depend on how much energy is required to be generated, generally up to 10kWp for a standard domestic installation in New Zealand.

The DC energy from the solar array can be stored in batteries and then converted to AC by an inverter as when required or you can convert the solar energy directly to AC with a grid connected inverter and feed any excess energy into the national grid for credits.

Working

Solar photovoltaic is an elegant technology which produces electricity from sunlight without moving parts.

In a photovoltaic cell, sunlight detaches electrons from their host silicon atoms. Tiny packets of light energy called photons are captured by electrons, and impart enough energy to kick the electron free of its host atom. Near the upper surface of the cell is a "one way membrane" called a pn-junction. The pn-junction is formed by diffusing tiny quantities of phosphorus to a depth of about one micrometre into a thin wafer of silicon.

When a free electron crosses the pn-junction it cannot easily return, causing a negative voltage to appear on the surface facing the sun (and a positive voltage on the rear surface). The front and rear surfaces can be connected together via an external circuit in order to extract current, voltage and power from the solar cell.

Solar cells are packaged behind glass to form photovoltaic modules, which have typical service lives of 20 to 40 years.

In many circumstances, photovoltaic modules mounted on building roofs can produce as much electricity as the building consumes. A typical module will generate about 200 kilowatt hours (AC)

per square meter per year, so a collector area of 25-50m^2 is needed to power a reasonably energy-efficient Australian house. Such a house exports more electricity to the grid during the day than it imports at night.

An additional 10m^2 is required to offset the annual greenhouse gas emissions of a fuel efficient car emitting 0.2kg of carbon dioxide per km and driving 10,000km per year.

PHOTOVOLTAIC EFFECT

Photovoltaic effect is a process in which two dissimilar materials in close contact produce an electrical voltage when struck by light or other radiant energy. Light striking crystals such as silicon or germanium, in which electrons are usually not free to move from atom to atom within the crystal, provides the energy needed to free some electrons from their bound condition. Free electrons cross the junction between two dissimilar crystals more easily in one direction than in the other, giving one side of the junction a negative charge and, therefore, a negative voltage with respect to the other side, just as one electrode of a battery has a negative voltage with respect to the other. The photovoltaic effect can continue to provide voltage and current as long as light continues to fall on the two materials. This current can be used to measure the brightness of the incident light or as a source of power in an electrical circuit, as in a solar power system.

The photovoltaic effect in a solar cell can be illustrated with an analogy to a child at a slide. Initially, both the electron and the child are in their respective "ground states." Next, the electron is lifted up to its excited state by consuming energy received from the incoming light, just as the child is lifted up to an "excited state" at the top of the slide by consuming chemical energy stored in his body. In both cases there is now energy available in the excited state that can be expended. In the absence of junction-forming materials, there is no incentive for excited, free electrons to move along a specific direction; they eventually fall back to the ground state. On the other hand, whenever two different materials are placed in contact, an electric field is generated along the contact. This is the so-called built-in field, and it exerts a force on free electrons, effectively "tilting" the electron states and forcing the excited free electrons into an external electrical load where their excess energy can be dissipated. The external load can be a simple resistor, or it can be any of a myriad of electrical or electronic devices ranging from motors to radios. Correspondingly, the child moves to the slide because of his desire for excitement. It is on the slide that the child dissipates his excess energy. Finally, when the excess energy is expended, both the electron and the child are back in the ground state, where they can begin the whole process over again. The motion of the electron, like that of the child, is in one direction, as can be seen from the figure. In short, the photovoltaic effect produces a direct current (DC)—one that flows constantly in only a single direction.

PHOTOELECTRIC EFFECT

The photoelectric effect is a quantum electronic phenomenon in which electrons are emitted from matter after the absorption of energy from electromagnetic radiation such as x-rays or visible light. The emitted electrons can be referred to as photoelectrons in this context. The effect is also termed the Hertz Effect, based on its discovery by Heinrich Hertz, although the term has generally fallen out of use.

Study of the photoelectric effect led to important steps in understanding the quantum nature of light and electrons and influenced the formation of the concept of wave–particle duality.

The term may also refer to the photoconductive effect (also known as photoconductivity or photoresistivitity), the photovoltaic effect, or the photoelectrochemical effect.

When a metallic surface is exposed to electromagnetic radiation above a certain threshold frequency, the light is absorbed and electrons are emitted. In 1902, Philipp Eduard Anton von Lenard observed that the energy of the emitted electrons increased with the frequency, or color, of the light. This was at odds with James Clerk Maxwell's wave theory of light, which predicted that the energy would be proportional to the intensity of the radiation. In 1905, Albert Einstein solved this paradox by describing light as composed of discrete quanta, now called photons, rather than continuous waves. Based upon Max Planck's theory of black-body radiation, Einstein theorized that the energy in each quantum of light was equal to the frequency multiplied by a constant, later called Planck's constant. A photon above a threshold frequency has the required energy to eject a single electron, creating the observed effect. This discovery led to the quantum revolution in physics and earned Einstein the Nobel Prize in 1921.

Explanation

The photons of the light beam have a characteristic energy determined by the frequency of the light. In the photoemission process, if an electron absorbs the energy of one photon and has more energy than the work function (the electron binding energy), it is ejected from the material. If the photon energy is too low, the electron is unable to escape the surface of the material. Increasing the intensity of the light beam increases the number of photons in the light beam, and thus increases the number of electrons emitted without increasing the energy that each electron possesses. Thus the energy of the emitted electrons does not depend on the intensity of the incoming light, but only on the energy of the individual photons.

Electrons can absorb energy from photons when irradiated, but they follow an "all or nothing" principle. All of the energy from one photon must be absorbed and used to liberate one electron from atomic binding, or the energy is re-emitted. If the photon energy is absorbed, some of the energy liberates the electron from the atom, and the rest contributes to the electron's kinetic energy as a free particle.

Experimental Results of the Photoelectric Emission

- For a given metal and frequency of incident radiation, the rate at which photoelectrons are ejected is directly proportional to the intensity of the incident light.
- For a given metal, there exists a certain minimum frequency of incident radiation below which no photoelectrons can be emitted. This frequency is called the threshold frequency.
- Above the threshold frequency, the maximum kinetic energy of the emitted photoelectron is independent of the intensity of the incident light but depends on the frequency of the incident light.

- The time lag between the incidence of radiation and the emission of a photoelectron is very small, less than 10^{-9} second.

Equations

In effect quantitatively using Einstein's method, the following equivalent equations are used:

Energy of photon = Energy needed to remove an electron + Kinetic energy of the emitted electron

Algebraically:

$$hf = \phi + E_{k_{max}}$$

where,

- h is Planck's constant,
- f is the frequency of the incident photon,
- $\phi = hf_0$ is the work function (sometimes denoted W instead), the minimum energy required to remove a delocalized electron from the surface of any given metal,
- $E_{k_{max}} = \frac{1}{2}mv_m^2$ is the maximum kinetic energy of ejected electrons,
- fo is the threshold frequency for the photoelectric effect to occur,
- m is the rest mass of the ejected electron, and
- υ_m is the speed of the ejected electron.

Since an emitted electron cannot have negative kinetic energy, the equation implies that if the photon's energy (hf) is less than the work function (ϕ), no electron will be emitted.

According to Einstein's special theory of relativity the relation between energy (E) and momentum (p) of a particle is $E = \sqrt{(pc)^2 + (mc^2)^2}$, where m is the rest mass of the particle and c is the velocity of light in a vacuum.

Three-step Model

The photoelectric effect in crystalline material is often decomposed into three steps:

- Inner photoelectric effect. The hole left behind can give rise to auger effect, which is visible even when the electron does not leave the material. In molecular solids photons are excited in this step and may be visible as lines in the final electron energy. The inner photoeffect has to be dipole allowed. The transition rules for atoms translate via the tight-binding model onto the crystal. They are similar in geometry to plasma oscillations in that they have to be transversal.
- Ballistic transport of half of the electrons to the surface. Some electrons are scattered.
- Electrons escape from the material at the surface.

In the three-step model, an electron can take multiple paths through these three steps. All paths can interfere in the sense of the path integral formulation. For surface states and molecules the three-step model does still make some sense as even most atoms have multiple electrons which can scatter the one electron leaving.

Uses and Effects

Photodiodes and Phototransistors

Solar cells (used in solar power) and light-sensitive diodes use a variant of the photoelectric effect, but not ejecting electrons out of the material. In semiconductors, light of even relatively low energy, such as visible photons, can kick electrons out of the valence band and into the higher-energy conduction band, where they can be harnessed, creating electric current at a voltage related to the bandgap energy.

Image Sensors

Video camera tubes in the early days of television used the photoelectric effect; newer variants used photoconductive rather than photoemissive materials.

Silicon image sensors, such as charge-coupled devices, widely used for photographic imaging, are based on a variant of the photoelectric effect, in which photons knock electrons out of the valence band of energy states in a semiconductor, but not out of the solid itself.

The Gold-leaf Electroscope

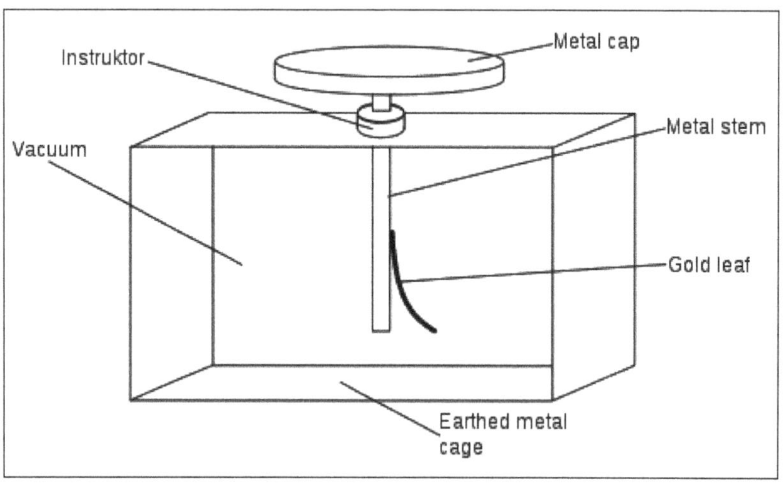

The gold leaf electroscope.

The electroscope is an important tool in illustrating the photoelectric effect. Let us say that the scope is negatively charged throughout. There is an excess of electrons and the leaf is separated from the stem. But if we then shine high-frequency light onto the cap, the scope discharges and the leaf will fall limp. This is because the frequency of the light shining on the cap is above the cap's threshold frequency. The photons in the light have enough energy to liberate electrons from the cap, reducing its negative charge. This will discharge a negatively charged electroscope and further charge a positive electroscope.

However, if the EM radiation hitting the metal cap does not have a high enough frequency, (its frequency is below the threshold value for the cap) then the leaf will never discharge, no matter how long one shines the low-frequency light at the cap.

Photoelectron Spectroscopy

Since the energy of the photoelectrons emitted is exactly the energy of the incident photon minus the material's work function or binding energy, the work function of a sample can be determined by bombarding it with a monochromatic X-ray source or UV source (typically a helium discharge lamp), and measuring the kinetic energy distribution of the electrons emitted.

Photoelectron spectroscopy is done in a high vacuum environment, since the electrons would be scattered by air.

A typical electron energy analyzer is a concentric hemispherical analyzer (CHA), which uses an electric field to divert electrons different amounts depending on their kinetic energies. For every element and core (atomic orbital) there will be a different binding energy. The many electrons created from each will then show up as spikes in the analyzer, and can be used to determine the elemental composition of the sample.

Spacecraft

The photoelectric effect will cause spacecraft exposed to sunlight to develop a positive charge. This can get up to the tens of volts. This can be a major problem, as other parts of the spacecraft in shadow develop a negative charge (up to several kilovolts) from nearby plasma, and the imbalance can discharge through delicate electrical components. The static charge created by the photoelectric effect is self-limiting, though, because a more highly-charged object gives up its electrons less easily.

Moon Dust

Light from the sun hitting lunar dust causes it to become charged through the photoelectric effect. The charged dust then repels itself and lifts off the surface of the Moon by electrostatic levitation. This manifests itself almost like an "atmosphere of dust," visible as a thin haze and blurring of distant features, and visible as a dim glow after the sun has set. This was first photographed by the Surveyor program probes in the 1960s. It is thought that the smallest particles are repelled up to kilometers high, and that the particles move in "fountains" as they charge and discharge.

Night Vision Devices

Photons hitting a gallium arsenide plate in night vision devices cause the ejection of photoelectrons due to the photoelectric effect. These are then amplified into a cascade of electrons that light up a phosphor screen.

References

- Solar-energy, science: britannica.com, Retrieved 31 March, 2019

- What-does-pv-or-photo-voltaic-mean: esolar.co.nz, Retrieved 14 July, 2019
- Explainer-what-is-photovoltaic-solar-energy-12924: theconversation.com, Retrieved 17 May, 2019
- Photovoltaic-effect, science: britannica.com, Retrieved 19 April, 2019
- Photoelectric-effect, entry: newworldencyclopedia.org, Retrieved 5 February, 2019

Photovoltaics: Concepts and Techniques

CHAPTER 2

Some of the common concepts of photovoltaics are solar radiation, solar irradiance, solar constant, solar cell efficiency, etc. A number of techniques are used in photovoltaics such as sun-free photovoltaics, polarizing organic photovoltaics and cadmium telluride photovoltaics. This chapter has been carefully written to provide an easy understanding of these concepts and techniques of photovoltaics.

SOLAR RADIATION

Solar Radiation is the electromagnetic and corpuscular radiation emitted by the sun. Solar radiation is the principal source of energy for most processes that occur on the earth.

The corpuscular emissions of the sun consist primarily of protons, which have speeds of 300–1,500 km/sec near the earth. The proton concentration in the vicinity of the earth is 5–80 ions/cm^3, but it increases with increasing solar activity and after large flares can reach 10^3 ions/cm3. During solar flares, particles with energies of 5×10^7 to $2 \times 10'^{o}$ electron volts are produced. Mainly protons, these energetic particles form the solar component of cosmic rays and partially account for the variations in the cosmic rays that reach the earth.

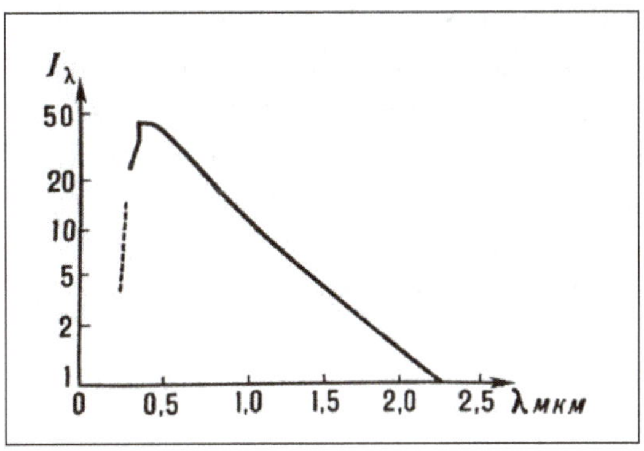

Dependence of radiated energy l_x on wavelength λ for the center of the solar disk; the unit of intensity is 10" ergs/(cm$^!$-sec-steradian).

The bulk of the electromagnetic radiation emitted by the sun is in the visible part of the spectrum. What is known as the solar constant is the amount of radiant energy received from the sun in 1 min on a surface 1 cm^2 in area perpendicular to the sun's rays and located outside the earth's atmosphere at the earth's mean distance from the sun. The solar constant is equal to 1.95 calories/(cm^2-min) and corresponds to a flux of 1.36 x 10^6ergs/(cm^2-sec). The emission of solar radiation is believed to increase at times of maximum solar activity. If, however, this increase does occur, it does not exceed a fraction of 1 percent.

The The radio-frequency radio- radiation emitted by the sun, or solar radio emission, is transmitted incompletely through the earth's atmosphere, since in the the radio-frequency radio- range the atmosphere is transparent only for wavelengths of a few millimeters to several meters. Solar radio emission is rather weak. It is measured in units of Φ = 10”^{22} watts/(m^2-sec-hertz) and varies from a few <t> to tens and hundreds of thousands of Φ as we move from the meter band, where frequencies are of the order of 10^8 hertz, to the millimeter band, where frequencies are of the order of 10” hertz. For a terrestrial observer, however, the sun, because of its relatively small distance from the earth, is the most powerful source of cosmic radiofrequency radiation. Solar radio emission consists of the thermal radio emission of the outer layers of the atmosphere of the quiet sun, a slowly varying component associated with sunspots and faculae, and sporadic radio emissions associated with solar activity. The sporadic radio-frequency radiation is often polarized, is more intense than the thermal emission, and varies rather rapidly. It includes noise storms and bursts. Bursts can be classified into five types, which differ both in frequency composition and in the character of the time dependence of the variations in intensity. Most bursts accompany solar flares.

The short-wavelength radiation of the sun is completely absorbed by the earth's atmosphere. Information on such radiation has been obtained by means of equipment carried by geophysical rockets, artificial earth satellites, and space probes. The intensity of the continuouso solar spectrum decreases abruptly at about 2,085 angstroms (A). The Fraunhofer lines vanish in the vicinity of 1,550 A. Although the= continuous spectrum can be traced to 1,000 A, beyond 1,500 A the spectrum consists principally of emission lines, such as the lines of hydrogen, ionized helium, and multiply ionized carbon, oxygen, and magnesium atoms. There are more than 200 emission lines in the ultraviolet region; the strongest is the resonance line of hydrogen (L_0) with a wavelength of 1,216 A. At the earth's orbit, the flux of short-wavelength radiation from the entire solar disk is 3–6 ergs/(cm^2-sec).

Solar X-radiation (wavelengths from 100 to 1 A) consists of continuous radiation and line emissions. The intensity of the Xradiation is strongly dependent on solar activity and varies from 0.13 erg/(cm^2-sec) to 1 erg/(cm^2-sec) at the earth's orbit. The X-ray spectrum becomes harder in years of maximum solar activity. During solar flares, the intensity of solar X-radiation is increased by a factor of several tens, and its hardness also increases. Although the ultraviolet radiation and Xradiation from the sun carry comparatively little energy—less than 15 ergs/(cm^2-sec) near the earth's orbit—such radiation has a very strong influence on the state of the upper layers of the earth's atmosphere. Solar gamma radiation has also been detected, but it remains insufficiently studied.

Basic Principles

Every location on Earth receives sunlight at least part of the year. The amount of solar radiation that reaches any one spot on the Earth's surface varies according to:

- Geographic location,
- Time of day,
- Season,
- Local landscape,
- Local weather.

Because the Earth is round, the sun strikes the surface at different angles, ranging from 0° (just above the horizon) to 90° (directly overhead). When the sun's rays are vertical, the Earth's surface gets all the energy possible. The more slanted the sun's rays are, the longer they travel through the atmosphere, becoming more scattered and diffuse. Because the Earth is round, the frigid polar regions never get a high sun, and because of the tilted axis of rotation, these areas receive no sun at all during part of the year.

The Earth revolves around the sun in an elliptical orbit and is closer to the sun during part of the year. When the sun is nearer the Earth, the Earth's surface receives a little more solar energy. The Earth is nearer the sun when it is summer in the southern hemisphere and winter in the northern hemisphere. However, the presence of vast oceans moderates the hotter summers and colder winters one would expect to see in the southern hemisphere as a result of this difference.

The 23.5° tilt in the Earth's axis of rotation is a more significant factor in determining the amount of sunlight striking the Earth at a particular location. Tilting results in longer days in the northern hemisphere from the spring (vernal) equinox to the fall (autumnal) equinox and longer days in the southern hemisphere during the other 6 months. Days and nights are both exactly 12 hours long on the equinoxes, which occur each year on or around March 23 and September 22.

Countries such as the United States, which lie in the middle latitudes, receive more solar energy in the summer not only because days are longer, but also because the sun is nearly overhead. The sun's rays are far more slanted during the shorter days of the winter months. Cities such as Denver, Colorado, (near 40° latitude) receive nearly three times more solar energy in June than they do in December.

The rotation of the Earth is also responsible for hourly variations in sunlight. In the early morning and late afternoon, the sun is low in the sky. Its rays travel further through the atmosphere than at noon, when the sun is at its highest point. On a clear day, the greatest amount of solar energy reaches a solar collector around solar noon.

Diffuse and Direct Solar Radiation

As sunlight passes through the atmosphere, some of it is absorbed, scattered, and reflected by:

- Air molecules,
- Water vapor,
- Clouds,
- Dust,
- Pollutants,
- Forest fires,
- Volcanoes.

This is called diffuse solar radiation. The solar radiation that reaches the Earth's surface without being diffused is called direct beam solar radiation. The sum of the diffuse and direct solar radiation is called global solar radiation. Atmospheric conditions can reduce direct beam radiation by 10% on clear, dry days and by 100% during thick, cloudy days.

Measurement

Scientists measure the amount of sunlight falling on specific locations at different times of the year. They then estimate the amount of sunlight falling on regions at the same latitude with similar climates. Measurements of solar energy are typically expressed as total radiation on a horizontal surface,or as total radiation on a surface tracking the sun.

Radiation data for solar electric (photovoltaic) systems are often represented as kilowatt-hours per square meter (kWh/m²). Direct estimates of solar energy may also be expressed as watts per square meter (W/m²).

Radiation data for solar water heating and space heating systems are usually represented in British thermal units per square foot (Btu/ft²).

SOLAR IRRADIANCE

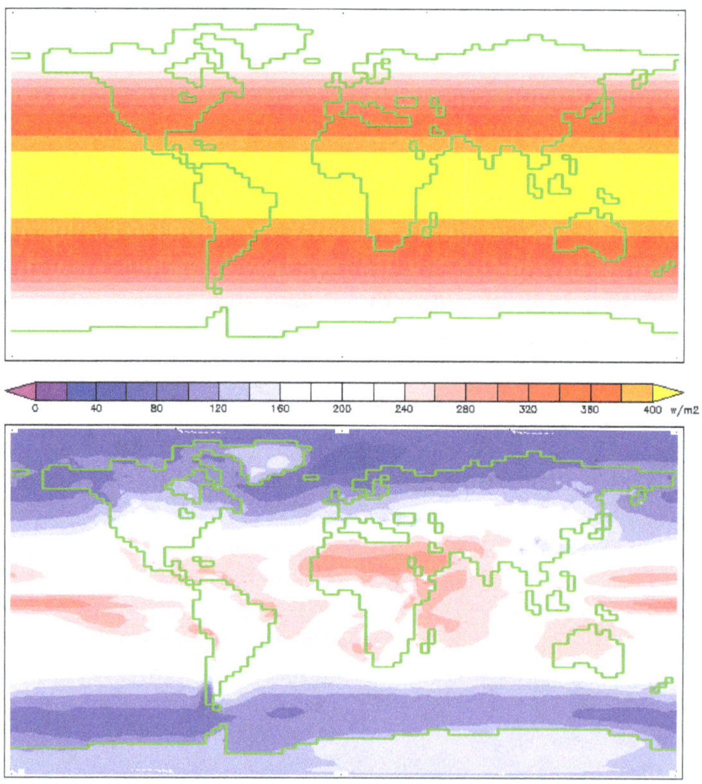

The shield effect of Earth's atmosphere on solar irradiation.

The top image is the annual mean solar irradiation (or insolation) at the top of Earth's atmosphere (TOA); the bottom image shows the annual insolation reaching the Earth's surface after passing through the atmosphere. Note that the two images use the same color scale.

Solar irradiance (SI) is the power per unit area (watt per square metre, W/m²), received from the Sun in the form of electromagnetic radiation as reported in the wavelength range of the measuring instrument.

Solar irradiance is often integrated over a given time period in order to report the radiant energy emitted into the surrounding environment (joule per square metre, J/m^2), during that time period. This integrated solar irradiance is called solar irradiation, solar exposure, solar insolation, or insolation.

Irradiance may be measured in space or at the Earth's surface after atmospheric absorption and scattering. Irradiance in space is a function of distance from the Sun, the solar cycle, and cross-cycle changes. Irradiance on the Earth's surface additionally depends on the tilt of the measuring surface, the height of the sun above the horizon, and atmospheric conditions. Solar irradiance affects plant metabolism and animal behavior.

The study and measurement of solar irradiance have several important applications, including the prediction of energy generation from solar power plants, the heating and cooling loads of buildings, and in climate modeling and weather forecasting.

Types

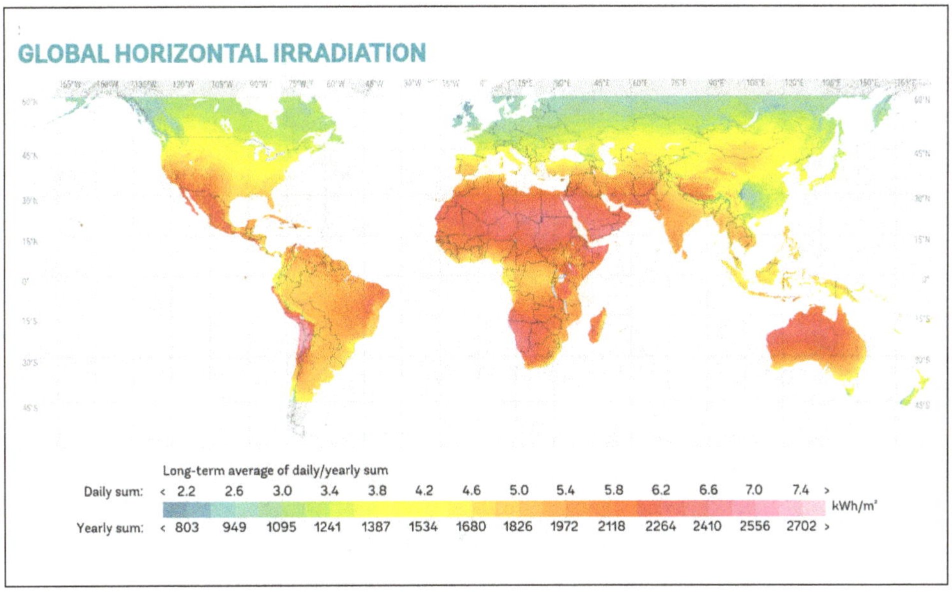

Global Map of Global Horizontal Radiation.

There are several measured types of solar irradiance:

- Total Solar Irradiance (TSI) is a measure of the solar power over all wavelengths per unit area incident on the Earth's upper atmosphere. It is measured perpendicular to the incoming sunlight. The solar constant is a conventional measure of mean TSI at a distance of one astronomical unit (AU).
- Direct Normal Irradiance (DNI), or beam radiation, is measured at the surface of the Earth at a given location with a surface element perpendicular to the Sun. It excludes diffuse solar radiation (radiation that is scattered or reflected by atmospheric components). Direct irradiance is equal to the extraterrestrial irradiance above the atmosphere minus the atmospheric losses due to absorption and scattering. Losses depend on time of day (length of light's path through the atmosphere depending on the solar elevation angle), cloud cover, moisture content and other contents. The irradiance above the atmosphere also varies with

time of year (because the distance to the sun varies), although this effect is generally less significant compared to the effect of losses on DNI.

- Diffuse Horizontal Irradiance (DHI), or Diffuse Sky Radiation is the radiation at the Earth's surface from light scattered by the atmosphere. It is measured on a horizontal surface with radiation coming from all points in the sky excluding circumsolar radiation (radiation coming from the sun disk). There would be almost no DHI in the absence of atmosphere.
- Global Horizontal Irradiance (GHI) is the total irradiance from the sun on a horizontal surface on Earth. It is the sum of direct irradiance (after accounting for the solar zenith angle of the sun z) and diffuse horizontal irradiance:

  $$\text{GHI} = \text{DHI} + \text{DNI} \times \cos(z)$$

- Global Normal Irradiance (GNI) is the total irradiance from the sun at the surface of Earth at a given location with a surface element perpendicular to the Sun.

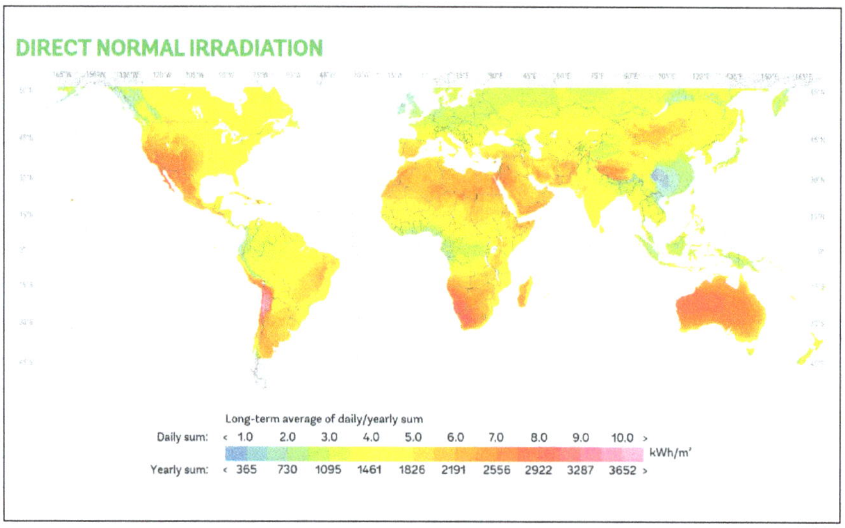

Global Map of Direct Normal Radiation.

Units

The SI unit of irradiance is watt per square metre (W/m^2, which may also be written Wm^{-2}).

An alternative unit of measure is the Langley (1 thermochemical calorie per square centimeter or 41,840 J/m^2) per unit time.

The solar energy industry uses watt-hour per square metre (Wh/m^2) per unit time. The relation to the SI unit is thus:

$$1\ kW/m^2 \times (24\ h/day) \times = (24\ kWh/m^2)/day$$

$$(24 kWh/m^2)/day * (365 days/year) = (8760\ kWh/m^2)/year.$$

Irradiation at the Top of the Atmosphere

The distribution of solar radiation at the top of the atmosphere is determined by Earth's sphericity and orbital parameters. This applies to any unidirectional beam incident to a rotating sphere. Insolation is essential for numerical weather prediction and understanding seasons and climate change. Application to ice ages is known as Milankovitch cycles.

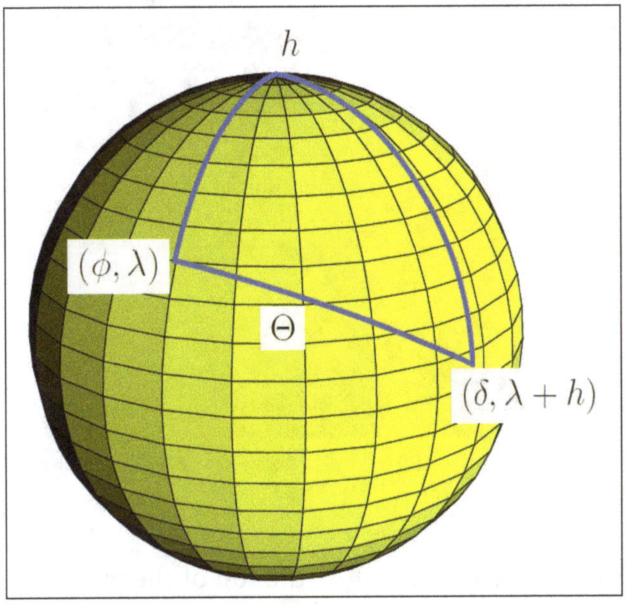

Spherical triangle for application of the spherical law of cosines for the calculation the solar zenith angle Θ for observer at latitude φ and longitude λ from knowledge of the hour angle h and solar declination δ. (δ is latitude of subsolar point, and h is relative longitude of subsolar point).

Distribution is based on a fundamental identity from spherical trigonometry, the spherical law of cosines:

$$\cos(c) = \cos(a)\cos(b) + \sin(a)\sin(b)\cos(C)$$

where *a*, *b* and *c* are arc lengths, in radians, of the sides of a spherical triangle. *C* is the angle in the vertex opposite the side which has arc length *c*. Applied to the calculation of solar zenith angle Θ, the following applies to the spherical law of cosines:

$$C = h$$

$$c = \Theta$$

$$a = \tfrac{1}{2}\pi - \phi$$

$$b = \tfrac{1}{2}\pi - \delta$$

$$\cos(\Theta) = \sin(\phi)\sin(\delta) + \cos(\phi)\cos(\delta)\cos(h)$$

This equation can be also derived from a more general formula:

$$\cos(\Theta) = \sin(\phi)\sin(\delta)\cos(\beta) + \sin(\delta)\cos(\phi)\sin(\beta)\cos(\gamma) + \cos(\phi)\cos(\delta)\cos(\beta)\cos(h) \\ - \cos(\delta)\sin(\phi)\sin(\beta)\cos(\gamma)\cos(h) - \cos(\delta)\sin(\beta)\sin(\gamma)\sin(h)$$

where β is an angle from the horizontal and γ is an azimuth angle.

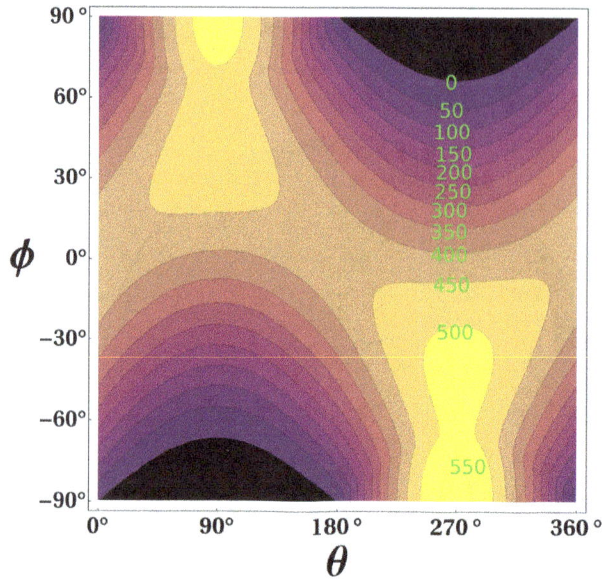

$\bar{Q}^{\text{day}}$, the theoretical daily-average irradiation at the top of the atmosphere, where θ is the polar angle of the Earth's orbit, and θ = 0 at the vernal equinox, and θ = 90° at the summer solstice; φ is the latitude of the Earth. The calculation assumed conditions appropriate for 2000 A.D.: a solar constant of S_0 = 1367 W m^{-2}, obliquity of ε = 23.4398°, longitude of perihelion of ϖ = 282.895°, eccentricity e = 0.016704. Contour labels (green) are in units of W m^{-2}.

The separation of Earth from the sun can be denoted R_E and the mean distance can be denoted R_0, approximately 1 astronomical unit (AU). The solar constant is denoted S_0. The solar flux density (insolation) onto a plane tangent to the sphere of the Earth, but above the bulk of the atmosphere (elevation 100 km or greater) is:

$$\begin{cases} \text{—}\cos(\Theta) & \cos(\Theta) > 0 \\ 0 & \cos(\Theta) \le 0 \end{cases}$$

The average of Q over a day is the average of Q over one rotation, or the hour angle progressing from $h = \pi$ to $h = -\pi$:

$$\bar{Q}^{\text{day}} = -\frac{1}{2\pi}\int_{\pi}^{-\pi} Q\,dh$$

Let h_0 be the hour angle when Q becomes positive. This could occur at sunrise when $\Theta = \frac{1}{2}\pi$,, or for h_0 as a solution of:

$$\sin(\phi)\sin(\delta) + \cos(\phi)\cos(\delta)\cos(h_o) = 0$$

or

$$\cos(h_o) = -\tan(\phi)\tan(\delta)$$

If tan(φ)tan(δ) > 1, then the sun does not set and the sun is already risen at $h = \pi$, so $h_o = \pi$. If tan(φ)tan(δ) < −1, the sun does not rise and $\bar{Q}^{\text{day}} = 0$.

$\frac{R_o^2}{R_E^2}$ is nearly constant over the course of a day, and can be taken outside the integral:

$$\begin{aligned}\int_{\pi}^{-\pi} Q dh &= \int_{h_o}^{-h_o} Q dh \\ &= S_o \frac{R_o^2}{R_E^2} \int_{h_o}^{-h_o} \cos(\Theta) dh \\ &= S_o \frac{R_o^2}{R_E^2} \left[h \sin(\phi)\sin(\delta) + \cos(\phi)\cos(\delta)\sin(h) \right]_{h=h_o}^{h=-h_o} \\ &= -2S_o \frac{R_o^2}{R_E^2} \left[h_o \sin(\phi)\sin(\delta) + \cos(\phi)\cos(\delta)\sin(h_o) \right]\end{aligned}$$

Therefore:

$$\bar{Q}^{\text{day}} = \frac{S_o}{\pi} \frac{R_o^2}{R_E^2} \left[h_o \sin(\phi)\sin(\delta) + \cos(\phi)\cos(\delta)\sin(h_o) \right]$$

Let θ be the conventional polar angle describing a planetary orbit. Let $\theta = 0$ at the vernal equinox. The declination δ as a function of orbital position is:

$$\delta = \varepsilon \sin(\theta)$$

where ε is the obliquity. The conventional longitude of perihelion ϖ is defined relative to the vernal equinox, so for the elliptical orbit:

$$R_E = \frac{R_o}{1 + e\cos(\theta - \varpi)}$$

or

$$\frac{R_o}{R_E} = 1 + e\cos(\theta - \varpi)$$

With knowledge of ϖ, ε and e from astrodynamical calculations and S_0 from a consensus of observations or theory, $\bar{Q}^{\text{day}}$ can be calculated for any latitude φ and θ. Because of the elliptical orbit, and as a consequence of Kepler's second law, θ does not progress uniformly with time. Nevertheless, $\theta = 0°$ is exactly the time of the vernal equinox, $\theta = 90°$ is exactly the time of the summer solstice, $\theta = 180°$ is exactly the time of the autumnal equinox and $\theta = 270°$ is exactly the time of the winter solstice.

A simplified equation for irradiance on a given day is:

$$Q = S_0 \left(1 + 0.034 \cos\left(2\pi \frac{n}{365.25} \right) \right)$$

where n is a number of a day of the year.

Variation

Total solar irradiance (TSI) changes slowly on decadal and longer timescales. The variation during solar cycle 21 was about 0.1% (peak-to-peak). In contrast to older reconstructions, most recent TSI reconstructions point to an increase of only about 0.05% to 0.1% between the Maunder Minimum and the present. Ultraviolet irradiance (EUV) varies by approximately 1.5 percent from solar maxima to minima, for 200 to 300 nm wavelengths. However, a proxy study estimated that UV has increased by 3.0% since the Maunder Minimum.

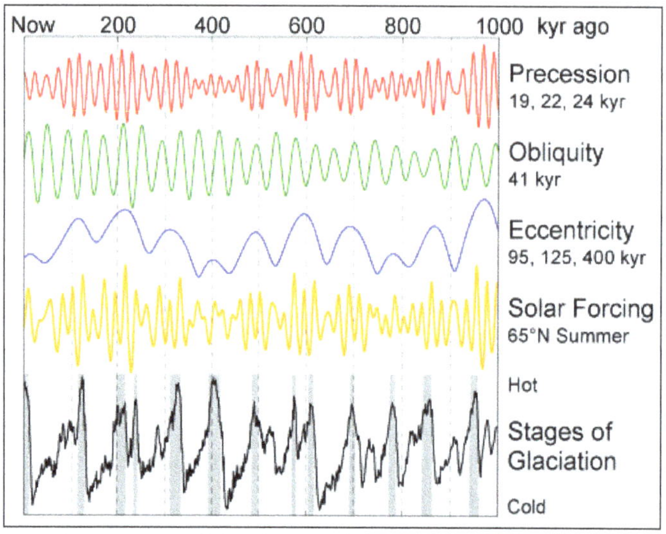

Variations in Earth's orbit, resulting changes in solar energy flux at high latitude, and the observed glacial cycles.

Some variations in insolation are not due to solar changes but rather due to the Earth moving between its perihelion and aphelion, or changes in the latitudinal distribution of radiation. These orbital changes or Milankovitch cycles have caused radiance variations of as much as 25% (locally; global average changes are much smaller) over long periods. The most recent significant event was an axial tilt of 24° during boreal summer near the *Holocene climatic optimum*. Obtaining a time series for a $\bar{Q}^{\mathrm{day}}$ for a particular time of year, and particular latitude, is a useful application in the theory of Milankovitch cycles. For example, at the summer solstice, the declination δ is equal to the obliquity ε. The distance from the sun is:

$$\frac{R_o}{R_E} = 1 + e\cos(\theta - \varpi) = 1 + e\cos(\tfrac{\pi}{2} - \varpi) = 1 + e\sin(\varpi)$$

For this summer solstice calculation, the role of the elliptical orbit is entirely contained within the important product , the precession index, whose variation dominates the variations in insolation at 65° N when eccentricity is large. For the next 100,000 years, with variations in eccentricity being relatively small, variations in obliquity dominate.

Measurement

The space-based TSI record comprises measurements from more than ten radiometers spanning three solar cycles. All modern TSI satellite instruments employ active cavity electrical substitution radiometry. This technique applies measured electrical heating to maintain an absorptive blackened cavity in thermal equilibrium while incident sunlight passes through a precision aperture of calibrated area. The aperture is modulated via a shutter. Accuracy uncertainties of $<0.01\%$ are required to detect long term solar irradiance variations, because expected changes are in the range 0.05 to 0.15 W/m^2 per century.

Intertemporal Calibration

In orbit, radiometric calibrations drift for reasons including solar degradation of the cavity, electronic degradation of the heater, surface degradation of the precision aperture and varying surface emissions and temperatures that alter thermal backgrounds. These calibrations require compensation to preserve consistent measurements.

For various reasons, the sources do not always agree. The Solar Radiation and Climate Experiment/Total Irradiance Measurement (SORCE/TIM) TSI values are lower than prior measurements by the Earth Radiometer Budget Experiment (ERBE) on the Earth Radiation Budget Satellite (ERBS), VIRGO on the Solar Heliospheric Observatory (SoHO) and the ACRIM instruments on the Solar Maximum Mission (SMM), Upper Atmosphere Research Satellite (UARS) and ACRIMSat. Pre-launch ground calibrations relied on component rather than system level measurements, since irradiance standards lacked absolute accuracies.

Measurement stability involves exposing different radiometer cavities to different accumulations of solar radiation to quantify exposure-dependent degradation effects. These effects are then compensated for in final data. Observation overlaps permits corrections for both absolute offsets and validation of instrumental drifts.

Uncertainties of individual observations exceed irradiance variability ($\sim 0.1\%$). Thus, instrument stability and measurement continuity are relied upon to compute real variations.

Long-term radiometer drifts can be mistaken for irradiance variations that can be misinterpreted as affecting climate. Examples include the issue of the irradiance increase between cycle minima in 1986 and 1996, evident only in the ACRIM composite (and not the model) and the low irradiance levels in the PMOD composite during the 2008 minimum.

Despite the fact that ACRIM I, ACRIM II, ACRIM III, VIRGO and TIM all track degradation with redundant cavities, notable and unexplained differences remain in irradiance and the modeled influences of sunspots and faculae.

Persistent Inconsistencies

Disagreement among overlapping observations indicates unresolved drifts that suggest the TSI record is not sufficiently stable to discern solar changes on decadal time scales. Only the ACRIM composite shows irradiance increasing by ~ 1 W/m^2 between 1986 and 1996; this change is also absent in the model.

Recommendations to resolve the instrument discrepancies include validating optical measurement accuracy by comparing ground-based instruments to laboratory references, such as those at National Institute of Science and Technology (NIST); NIST validation of aperture area calibrations uses spares from each instrument; and applying diffraction corrections from the view-limiting aperture.

For ACRIM, NIST determined that diffraction from the view-limiting aperture contributes a 0.13% signal not accounted for in the three ACRIM instruments. This correction lowers the reported ACRIM values, bringing ACRIM closer to TIM. In ACRIM and all other instruments but TIM, the aperture is deep inside the instrument, with a larger view-limiting aperture at the front. Depending on edge imperfections this can directly scatter light into the cavity. This design admits into the front part of the instrument two to three times the amount of light intended to be measured; if not completely absorbed or scattered, this additional light produces erroneously high signals. In contrast, TIM's design places the precision aperture at the front so that only desired light enters.

Variations from other sources likely include an annual systematics in the ACRIM III data that is nearly in phase with the Sun-Earth distance and 90-day spikes in the VIRGO data coincident with SoHO spacecraft maneuvers that were most apparent during the 2008 solar minimum.

TSI Radiometer Facility

TIM's high absolute accuracy creates new opportunities for measuring climate variables. TSI Radiometer Facility (TRF) is a cryogenic radiometer that operates in a vacuum with controlled light sources. L-1 Standards and Technology (LASP) designed and built the system, completed in 2008. It was calibrated for optical power against the NIST Primary Optical Watt Radiometer, a cryogenic radiometer that maintains the NIST radiant power scale to an uncertainty of 0.02% (1σ). As of 2011 TRF was the only facility that approached the desired <0.01% uncertainty for pre-launch validation of solar radiometers measuring irradiance (rather than merely optical power) at solar power levels and under vacuum conditions.

TRF encloses both the reference radiometer and the instrument under test in a common vacuum system that contains a stationary, spatially uniform illuminating beam. A precision aperture with area calibrated to 0.0031% (1σ) determines the beam's measured portion. The test instrument's precision aperture is positioned in the same location, without optically altering the beam, for direct comparison to the reference. Variable beam power provides linearity diagnostics, and variable beam diameter diagnoses scattering from different instrument components.

The Glory/TIM and PICARD/PREMOS flight instrument absolute scales are now traceable to the TRF in both optical power and irradiance. The resulting high accuracy reduces the consequences of any future gap in the solar irradiance record.

Difference Relative to TRF						
Instrument	Irradiance: View-Limiting Aperture Overfilled	Irradiance: Precision Aperture Overfilled	Difference Attributable To Scatter Error	Measured Optical Power Error	Residual Irradiance Agreement	Uncertainty
SORCE/TIM ground	NA	−0.037%	NA	−0.037%	0.000%	0.032%
Glory/TIM flight	NA	−0.012%	NA	−0.029%	0.017%	0.020%

PREMOS-1 ground	−0.005%	−0.104%	0.098%	−0.049%	−0.104%	~0.038%
PREMOS-3 flight	0.642%	0.605%	0.037%	0.631%	−0.026%	~0.027%
VIRGO-2 ground	0.897%	0.743%	0.154%	0.730%	0.013%	~0.025%

2011 Reassessment

The most probable value of TSI representative of solar minimum is 1360.9 ± 0.5 W/m^2, lower than the earlier accepted value of 1365.4 ± 1.3 W/m^2, established in the 1990s. The new value came from SORCE/TIM and radiometric laboratory tests. Scattered light is a primary cause of the higher irradiance values measured by earlier satellites in which the precision aperture is located behind a larger, view-limiting aperture. The TIM uses a view-limiting aperture that is smaller than precision aperture that precludes this spurious signal. The new estimate is from better measurement rather than a change in solar output.

A regression model-based split of the relative proportion of sunspot and facular influences from SORCE/TIM data accounts for 92% of observed variance and tracks the observed trends to within TIM's stability band. This agreement provides further evidence that TSI variations are primarily due to solar surface magnetic activity.

Instrument inaccuracies add a significant uncertainty in determining Earth's energy balance. The energy imbalance has been variously measured (during a deep solar minimum of 2005–2010) to be +0.58 ± 0.15 W/m^2), +0.60 ± 0.17 W/m^2 and +0.85 W/m^2. Estimates from space-based measurements range from +3 to 7 W/m^2. SORCE/TIM's lower TSI value reduces this discrepancy by 1 W/m^2. This difference between the new lower TIM value and earlier TSI measurements corresponds to a climate forcing of −0.8 W/m^2, which is comparable to the energy imbalance.

2014 Reassessment

In 2014 a new ACRIM composite was developed using the updated ACRIM3 record. It added corrections for scattering and diffraction revealed during recent testing at TRF and two algorithm updates. The algorithm updates more accurately account for instrument thermal behavior and parsing of shutter cycle data. These corrected a component of the quasi-annual spurious signal and increased the signal to noise ratio, respectively. The net effect of these corrections decreased the average ACRIM3 TSI value without affecting the trending in the ACRIM Composite TSI.

Differences between ACRIM and PMOD TSI composites are evident, but the most significant is the solar minimum-to-minimum trends during solar cycles 21-23. ACRIM found an increase of +0.037%/decade from 1980 to 2000 and a decrease thereafter. PMOD instead presents a steady decrease since 1978. Significant differences can also be seen during the peak of solar cycles 21 and 22. These arise from the fact that ACRIM uses the original TSI results published by the satellite experiment teams while PMOD significantly modifies some results to conform them to specific TSI proxy models. The implications of increasing TSI during the global warming of the last two decades of the 20th century are that solar forcing may be a marginally larger factor in climate change than represented in the CMIP5 general circulation climate models.

Irradiance on Earth's Surface

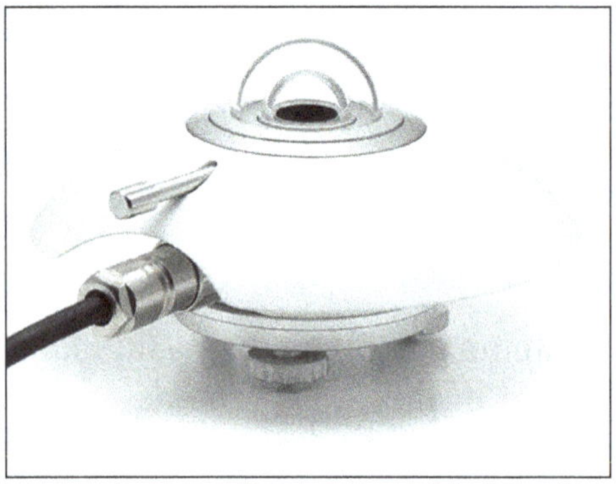

A pyranometer, used to measure global irradiance.

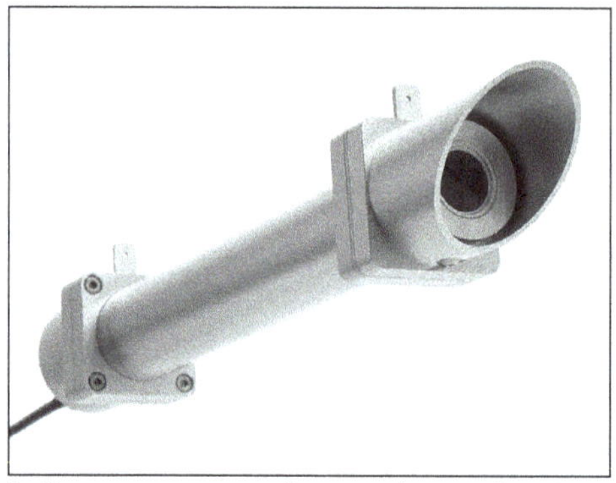

A pyrheliometer, mounted on a solar tracker, is used to measure Direct Normal Irradiance (or beam irradiance).

Average annual solar radiation arriving at the top of the Earth's atmosphere is roughly 1361 W/m². The Sun's rays are attenuated as they pass through the atmosphere, leaving maximum normal surface irradiance at approximately 1000 W /m² at sea level on a clear day. When 1361 W/m² is arriving above the atmosphere (when the sun is at the zenith in a cloudless sky), direct sun is about 1050 W/m², and global radiation on a horizontal surface at ground level is about 1120 W/m². The latter figure includes radiation scattered or reemitted by atmosphere and surroundings. The actual figure varies with the Sun's angle and atmospheric circumstances. Ignoring clouds, the daily average insolation for the Earth is approximately 6 kWh/m² = 21.6 MJ/m².

The output of, for example, a photovoltaic panel, partly depends on the angle of the sun relative to the panel. One Sun is a unit of power flux, not a standard value for actual insolation. Sometimes this unit is referred to as a Sol.

Absorption and Reflection

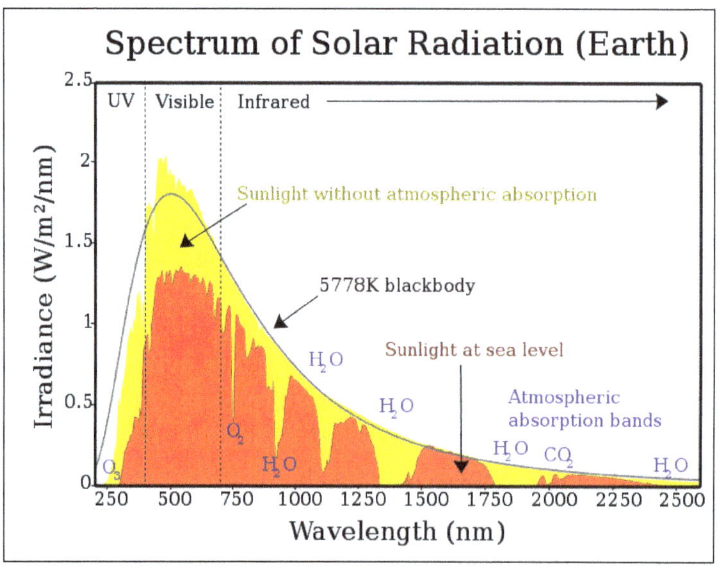

Part of the radiation reaching an object is absorbed and the remainder reflected. Usually the absorbed radiation is converted to thermal energy, increasing the object's temperature. Manmade or natural systems, however, can convert part of the absorbed radiation into another form such as electricity or chemical bonds, as in the case of photovoltaic cells or plants. The proportion of reflected radiation is the object's reflectivity or albedo.

Projection Effect

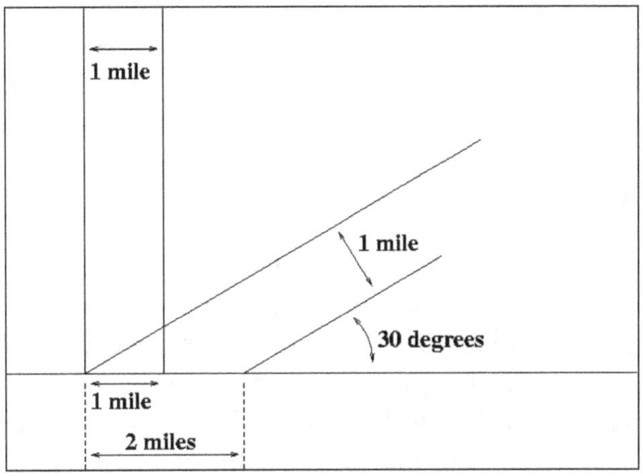

Projection effect: One sunbeam one mile wide shines on the ground at a 90° angle, and another at a 30° angle. The oblique sunbeam distributes its light energy over twice as much area.

Insolation onto a surface is largest when the surface directly faces (is normal to) the sun. As the angle between the surface and the Sun moves from normal, the insolation is reduced in proportion to the angle's cosine.

In the figure, the angle shown is between the ground and the sunbeam rather than between the vertical direction and the sunbeam; hence the sine rather than the cosine is appropriate. A sunbeam one mile (1.6 km) wide arrives from directly overhead, and another at a 30° angle to the horizontal. The sine of a 30° angle is 1/2, whereas the sine of a 90° angle is 1. Therefore, the angled sunbeam spreads the light over twice the area. Consequently, half as much light falls on each square mile.

This 'projection effect' is the main reason why Earth's polar regions are much colder than equatorial regions. On an annual average the poles receive less insolation than does the equator, because the poles are always angled more away from the sun than the tropics, and moreover receive no insolation at all for the six months of their respective winters.

Absorption Effect

At a lower angle the light must also travel through more atmosphere. This attenuates it (by absorption and scattering) further reducing insolation at the surface.

Attenuation is governed by the Beer-Lambert Law, namely that the transmittance or fraction of insolation reaching the surface decreases exponentially in the optical depth or absorbance (the two notions differing only by a constant factor of $ln(10) = 2.303$) of the path of insolation through the atmosphere. For any given short length of the path the optical depth is proportional to the quantity of absorbers and scatterers along that length, typically increasing with decreasing altitude. The optical depth of the whole path is then the integral (sum) of those optical depths along the path.

When the density of absorbers is layered, that is, depends much more on vertical than horizontal position in the atmosphere, to a good approximation the optical depth is inversely proportional to the projection effect, that is, to the cosine of the zenith angle. Since transmittance decreases exponentially with increasing optical depth, as the sun approaches the horizon there comes a point when absorption dominates projection for the rest of the day. With a relatively high level of absorbers this can be a considerable portion of the late afternoon, and likewise of the early morning. Conversely in the (hypothetical) total absence of absorption the optical depth remains zero at all altitudes of the sun, that is, transmittance remains 1, and so only the projection effect applies.

Solar Potential Maps

Assessment and mapping of solar potential at the global, regional and country levels has been the subject of significant academic and commercial interest. One of the earliest attempts to carry out comprehensive mapping of solar potential for individual countries was the Solar & Wind Resource Assessment (SWERA) project, funded by the United Nations Environment Program and carried out by the US National Renewable Energy Laboratory. Other examples include global mapping by the National Aeronautics and Space Administration and other similar institutes, many of which are available on the Global Atlas for Renewable Energy provided by the International Renewable Energy Agency. A number of commercial firms now exist to provide solar resource data to solar power developers, including 3E, Clean Power Research, Solargis, Vaisala (previously 3Tier), and Vortex, and these firms have often provided solar potential maps for free. In January 2017 the Global Solar Atlas was launched by the World Bank, using data provided by Solargis, to provide a single source for high quality solar data, maps, and GIS layers covering all countries.

Maps of GHI potential by region and country (Colors are not consistent across maps):

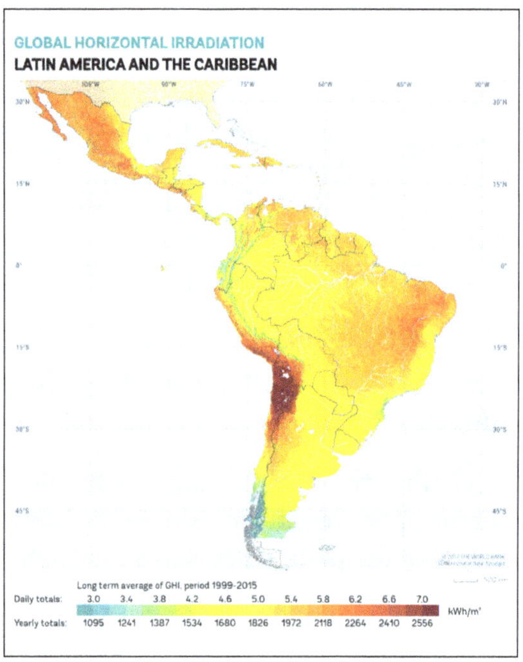

Latin America and Caribbean.

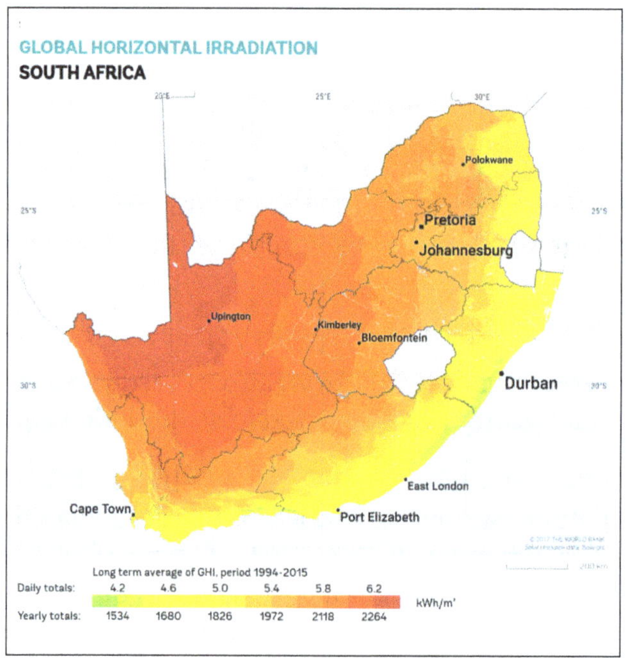

South Africa.

Applications

Solar Power

Conversion factor (multiply top row by factor to obtain side column)					
	W/m^2	kW·h/(m^2·day)	Sun hours/day	kWh/(m^2·y)	kWh/(kWp·y)
W/m^2	1	41.66666	41.66666	0.1140796	0.1521061
kW·h/(m^2·day)	0.024	1	1	0.0027379	0.0036505
sun hours/day	0.024	1	1	0.0027379	0.0036505
kWh/(m^2·y)	8.765813	365.2422	365.2422	1	1.333333
kWh/(kWp·y)	6.574360	273.9316	273.9316	0.75	1

Solar irradiation figures are used to plan the deployment of solar power systems. In many countries the figures can be obtained from an insolation map or from insolation tables that reflect data over the prior 30–50 years. Different solar power technologies are able to use different component of the total irradiation. While solar photovoltaics panels are able to convert to electricity both direct irradiation and diffuse irradiation, concentrated solar power is only able to operate efficiently with direct irradiation, thus making these systems suitable only in locations with relatively low cloud cover.

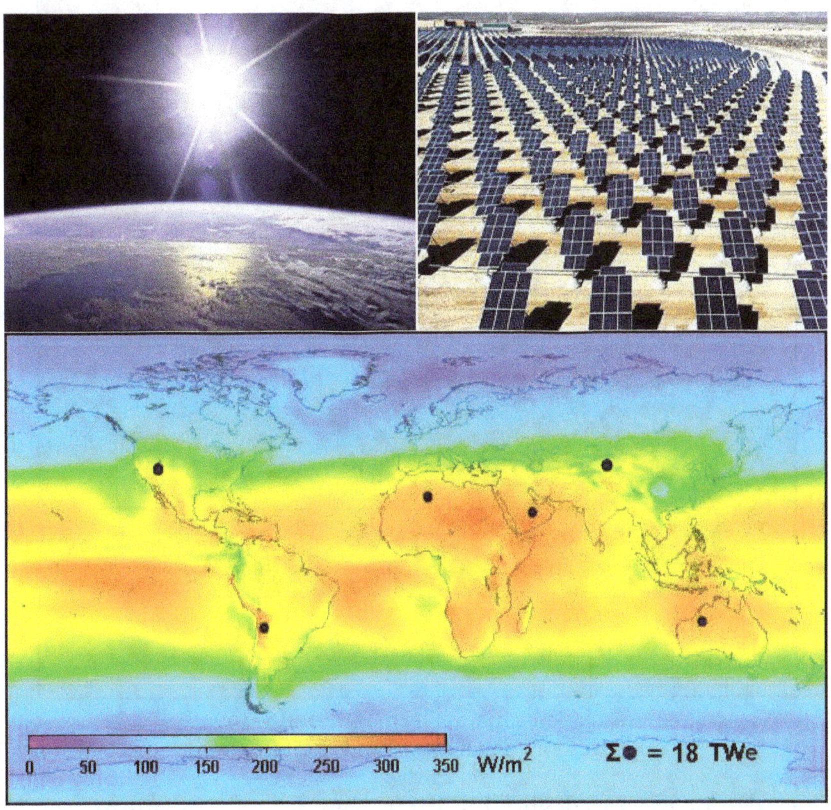

Sunlight carries radiant energy in the wavelengths of visible light.
Radiant energy may be developed for solar power generation.

Because solar collectors panels are almost always mounted at an angle towards the sun, insolation must be adjusted to prevent estimates that are inaccurately low for winter and inaccurately high for summer.

This also means that the amount of sun falling on a solar panel at high latitude is not as low compared to one at the equator as would appear from just considering insolation on a horizontal surface.

Photovoltaic panels are rated under standard conditions to determine the Wp (watt peak) rating, which can then be used with insolation to determine the expected output, adjusted by factors such as tilt, tracking and shading (which can be included to create the installed Wp rating). Insolation values range from 800 to 950 kWh/(kWp·y) in Norway to up to 2,900 kWh/(kWp·y) in Australia.

Buildings

In construction, insolation is an important consideration when designing a building for a particular site.

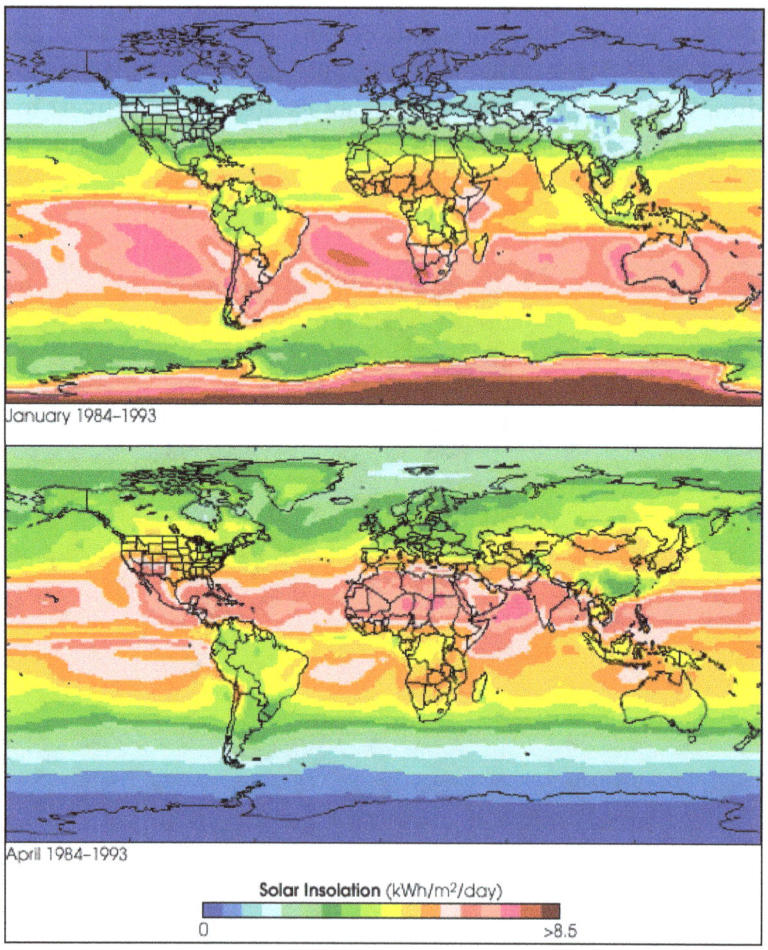

Insolation variation by month; 1984–1993 averages for January (top) and April (bottom).

The projection effect can be used to design buildings that are cool in summer and warm in winter, by providing vertical windows on the equator-facing side of the building (the south face in the northern hemisphere, or the north face in the southern hemisphere): this maximizes insolation in the winter months when the Sun is low in the sky and minimizes it in the summer when the Sun is high. (The Sun's north/south path through the sky spans 47 degrees through the year).

Civil Engineering

In civil engineering and hydrology, numerical models of snowmelt runoff use observations of insolation. This permits estimation of the rate at which water is released from a melting snowpack. Field measurement is accomplished using a pyranometer.

Climate Research

Irradiance plays a part in climate modeling and weather forecasting. A non-zero average global net radiation at the top of the atmosphere is indicative of Earth's thermal disequilibrium as imposed by climate forcing.

The impact of the lower 2014 TSI value on climate models is unknown. A few tenths of a percent change in the absolute TSI level is typically considered to be of minimal consequence for climate simulations. The new measurements require climate model parameter adjustments.

Experiments with GISS Model 3 investigated the sensitivity of model performance to the TSI absolute value during present and pre-industrial epochs, and describe, for example, how the irradiance reduction is partitioned between the atmosphere and surface and the effects on outgoing radiation.

Assessing the impact of long-term irradiance changes on climate requires greater instrument stability combined with reliable global surface temperature observations to quantify climate response processes to radiative forcing on decadal time scales. The observed 0.1% irradiance increase imparts 0.22 W/m^2 climate forcing, which suggests a transient climate response of 0.6 °C per W/m^2. This response is larger by a factor of 2 or more than in the IPCC-assessed 2008 models, possibly appearing in the models' heat uptake by the ocean.

Space Travel

Insolation is the primary variable affecting equilibrium temperature in spacecraft design and planetology.

Solar activity and irradiance measurement is a concern for space travel. For example, the American space agency, NASA, launched its Solar Radiation and Climate Experiment (SORCE) satellite with Solar Irradiance Monitors.

SOLAR CONSTANT

The solar constant (G_{SC}) is a flux density measuring mean solar electromagnetic radiation (solar irradiance) per unit area. It is measured on a surface perpendicular to the rays, one astronomical unit (AU) from the Sun (roughly the distance from the Sun to the Earth).

The solar constant includes all types of solar radiation, not just the visible light. It is measured by satellite as being 1.361 kilowatts per square meter (kW/m^2) at solar minimum and approximately 0.1% greater (roughly 1.362 kW/m^2) at solar maximum.

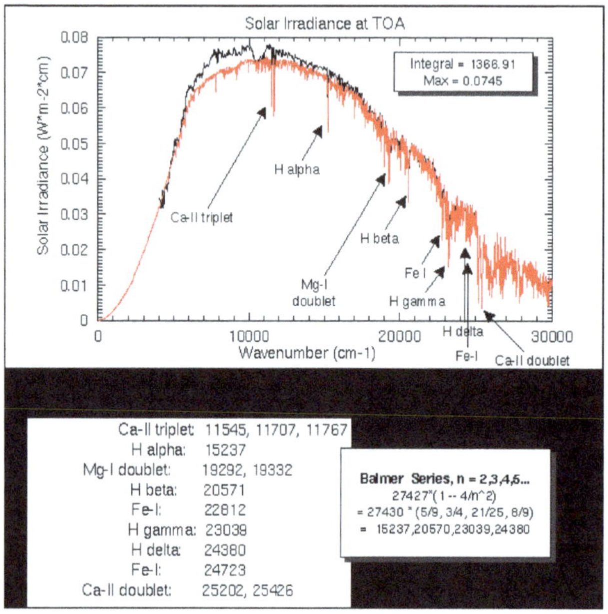

Solar irradiance spectrum at top of atmosphere, on a linear scale and plotted against wavenumber.

The solar "constant" is not a physical constant in the modern CODATA scientific sense; that is, it is not like the Planck constant or the speed of light which are absolutely constant in physics. The solar constant is an average of a varying value. In the past 400 years it has varied less than 0.2 percent. Billions of years ago, it was significantly lower.

This constant is used in the calculation of radiation pressure, which aids in the calculation of a force on a solar sail.

Calculation

Solar irradiance is measured by satellites above Earth's atmosphere, and is then adjusted using the inverse square law to infer the magnitude of solar irradiance at one Astronomical Unit (AU) to evaluate the solar constant. The approximate average value cited, 1.3608 ± 0.0005 kW/m², which is 81.65 kJ/m² per minute, is equivalent to approximately 1.951 calories per minute per square centimeter, or 1.951 langleys per minute.

Solar output is nearly, but not quite, constant. Variations in total solar irradiance (TSI) were small and difficult to detect accurately with technology available before the satellite era (±2% in 1954). Total solar output is now measured as varying (over the last three 11-year sunspot cycles) by approximately 0.

Relationship to other Measurements

Solar Irradiance

The actual direct solar irradiance at the top of the atmosphere fluctuates by about 6.9% during a year (from 1.412 kW/m² in early January to 1.321 kW/m² in early July) due to the Earth's varying distance from the Sun, and typically by much less than 0.1% from day to day. Thus, for the whole Earth (which has a cross section of 127,400,000 km²), the power is 1.730×10^{17} W (or 173,000 terawatts), plus or minus 3.5% (half the approximately 6.9% annual range). The solar constant does not remain constant over long periods of time, but over a year the solar constant varies much less than the solar irradiance measured at the top of the atmosphere. This is because the solar constant is evaluated at a fixed distance of 1 Astronomical Unit (AU) while the solar irradiance will be affected by the eccentricity of the Earth's orbit. Its distance to the Sun varies annually between $147.1 \cdot 10^6$ km at perihelion and $152.1 \cdot 10^6$ km at aphelion.

The Earth receives a total amount of radiation determined by its cross section ($\pi \cdot R_E^2$), but as it rotates this energy is distributed across the entire surface area ($4 \cdot \pi \cdot R_E^2$). Hence the average incoming solar radiation, taking into account the angle at which the rays strike and that at any one moment half the planet does not receive any solar radiation, is one-fourth the solar constant (approximately 340 W/m^2). The amount reaching the Earth's surface (as insolation) is further reduced by atmospheric attenuation, which varies. At any given moment, the amount of solar radiation received at a location on the Earth's surface depends on the state of the atmosphere, the location's latitude, and the time of day.

Apparent Magnitude

The solar constant includes all wavelengths of solar electromagnetic radiation, not just the visible light. It is positively correlated with the apparent magnitude of the Sun which is −26.8. The solar constant and the magnitude of the Sun are two methods of describing the apparent brightness of the Sun, though the magnitude is based on the Sun's visual output only.

The Sun's Total Radiation

The angular diameter of the Earth as seen from the Sun is approximately 1/11,700 radians (about 18 arcseconds), meaning the solid angle of the Earth as seen from the Sun is approximately 1/175,000,000 of a steradian. Thus the Sun emits about 2.2 billion times the amount of radiation that is caught by Earth, in other words about 3.846×10^{26} watts.

Past Variations in Solar Irradiance

Space-based observations of solar irradiance started in 1978. These measurements show that the solar constant is not constant. It varies with the 11-year sunspot solar cycle. When going further back in time, one has to rely on irradiance reconstructions, using sunspots for the past 400 years or cosmogenic radionuclides for going back 10,000 years. Such reconstructions show that solar irradiance varies with distinct periodicities. These cycles are: 11 years (Schwabe), 88 years (Gleisberg cycle), 208 years (DeVries cycle) and 1,000 years (Eddy cycle).

Over billions of years, the Sun is gradually expanding, and emitting more energy from the resultant larger surface area. The unsolved question of how to account for the clear geological evidence of liquid water on the Earth billions of years ago, at a time when the sun's luminosity was only 70% of its current value, is known as the faint young Sun paradox.

Variations due to Atmospheric Conditions

At most about 75% of the solar energy actually reaches the earth's surface, as even with a cloudless sky it is partially reflected and absorbed by the atmosphere. Even light cirrus clouds reduce this to

50%, stronger cirrus clouds to 40%. Thus the solar energy arriving at the surface can vary from 550 W/m^2 with cirrus clouds to 1025 W/m^2 with a clear sky.

SOLAR CELL EFFICIENCY

Solar cell efficiency refers to the portion of energy in the form of sunlight that can be converted via photovoltaics into electricity by the solar cell.

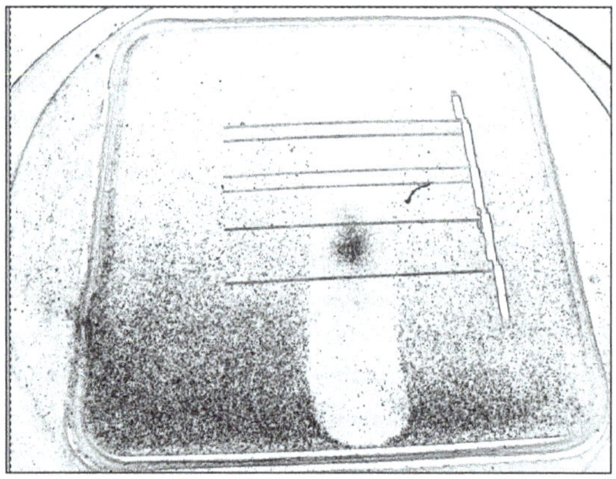

Dust often accumulates on the glass of solar modules - highlighted in this negative image as black dots - which reduces the amount of light admitted to the solar cells.

The efficiency of the solar cells used in a photovoltaic system, in combination with latitude and climate, determines the annual energy output of the system. For example, a solar panel with 20% efficiency and an area of 1 m^2 will produce 200 W at Standard Test Conditions, but it can produce more when the sun is high in the sky and will produce less in cloudy conditions or when the sun is low in the sky. In central Colorado, which receives annual insolation of 2000 $kWh/m^2/year$, such a panel can be expected to produce 400 kWh of energy per year. However, in Michigan, which receives only 1400 $kWh/m^2/year$, annual energy yield will drop to 280 kWh for the same panel. At more northerly European latitudes, yields are significantly lower: 175 kWh annual energy yield in southern England.

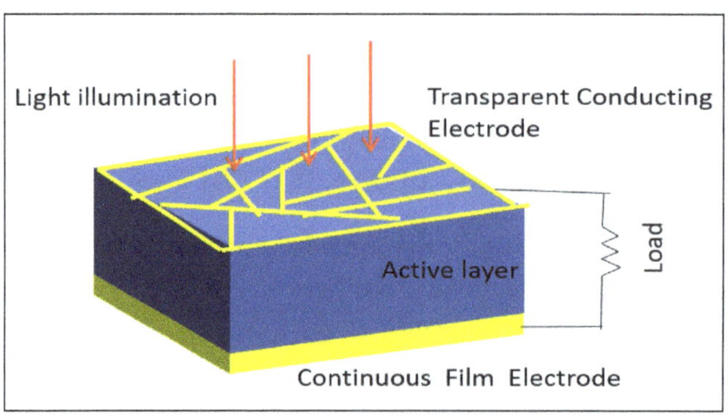

Schematic of charge collection by solar cells.

Light transmits through transparent conducting electrode creating electron hole pairs, which are collected by both the electrodes. The absorption and collection efficiencies of a solar cell depend on the design of transparent conductors and active layer thickness.

Several factors affect a cell's conversion efficiency value, including its reflectance efficiency, thermodynamic efficiency, charge carrier separation efficiency, charge carrier collection efficiency and conduction efficiency values. Because these parameters can be difficult to measure directly, other parameters are measured instead, including quantum efficiency, open-circuit voltage (V_{OC}) ratio, and § Fill factor. Reflectance losses are accounted for by the quantum efficiency value, as they affect "external quantum efficiency." Recombination losses are accounted for by the quantum efficiency, V_{OC} ratio, and fill factor values. Resistive losses are predominantly accounted for by the fill factor value, but also contribute to the quantum efficiency and V_{OC} ratio values.

In 2019, the world record for solar cell efficiency at 47.1% was achieved by using multi-junction concentrator solar cells, developed at National Renewable Energy Laboratory, Golden, Colorado, USA. This is above the standard rating of 37.0% for polycrystalline photovoltaic or thin-film solar cells.

Factors Affecting Energy Conversion Efficiency

Thermodynamic Efficiency Limit and Infinite-stack Limit

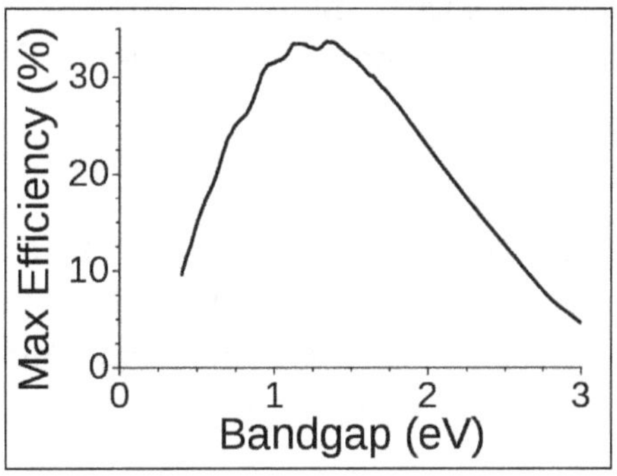

The Shockley–Queisser limit for the efficiency of a single-junction solar cell under unconcentrated sunlight at 273 K. This calculated curve uses actual solar spectrum data, and therefore the curve is wiggly from IR absorption bands in the atmosphere. This efficiency limit of ~34% can be exceeded by multijunction solar cells.

If one has a source of heat at temperature T_s and cooler heat sink at temperature T_c, the maximum theoretically possible value for the ratio of work (or electric power) obtained to heat supplied is $1\text{-}T_c/T_s$, given by a Carnot heat engine. If we take 6000 K for the temperature of the sun and 300 K for ambient conditions on earth, this comes to 95%. In 1981, Alexis de Vos and Herman Pauwels showed that this is achievable with a stack of an infinite number of cells with band gaps ranging from infinity (the first cells encountered by the incoming photons) to zero, with a voltage in each cell very close to the open-circuit voltage, equal to 95% of the band gap of that cell, and with 6000 K blackbody radiation coming from all directions. However, the 95% efficiency thereby achieved means that the electric power is 95% of the net amount of light absorbed – the stack emits radiation as it has non-zero temperature, and this radiation has to be subtracted from the incoming radiation when calculating the amount of heat being transferred and the efficiency. They also considered the more relevant problem of maximizing the power output for a stack being illuminated from all directions by 6000 K blackbody radiation. In this case, the voltages must be lowered to less than 95% of the band gap (the percentage is not constant over all the cells). The maximum theoretical efficiency calculated is 86.8% for a stack of an infinite number of cells, using the incoming concentrated sunlight radiation. When the incoming radiation comes only from an area of the sky the size of the sun, the efficiency limit drops to 68.7%.

Ultimate Efficiency

Normal photovoltaic systems however have only one p–n junction and are therefore subject to a lower efficiency limit, called the "ultimate efficiency" by Shockley and Queisser. Photons with an energy below the band gap of the absorber material cannot generate an electron-hole pair, so their energy is not converted to useful output, and only generates heat if absorbed. For photons with an energy above the band gap energy, only a fraction of the energy above the band gap can be converted to useful output. When a photon of greater energy is absorbed, the excess energy above the band gap is converted to kinetic energy of the carrier combination. The excess kinetic energy is converted to heat through phonon interactions as the kinetic energy of the carriers slows to equilibrium velocity. Traditional single-junction cells with an optimal band gap for the solar spectrum have a maximum theoretical efficiency of 33.16%, the Shockley–Queisser limit .

Solar cells with multiple band gap absorber materials improve efficiency by dividing the solar spectrum into smaller bins where the thermodynamic efficiency limit is higher for each bin.

Quantum Efficiency

As described above, when a photon is absorbed by a solar cell it can produce an electron-hole pair. One of the carriers may reach the p–n junction and contribute to the current produced by the solar cell; such a carrier is said to be *collected*. Or, the carriers recombine with no net contribution to cell current.

Quantum efficiency refers to the percentage of photons that are converted to electric current (i.e., collected carriers) when the cell is operated under short circuit conditions. The "external" quantum efficiency of a silicon solar cell includes the effect of optical losses such as transmission and reflection.

In particular, some measures can be taken to reduce these losses. The reflection losses, which can account for up to 10% of the total incident energy, can be dramatically decreased using a technique called texturization, a light trapping method that modifies the average light path.

Quantum efficiency is most usefully expressed as a spectral measurement (that is, as a function of photon wavelength or energy). Since some wavelengths are absorbed more effectively than others, spectral measurements of quantum efficiency can yield valuable information about the quality of the semiconductor bulk and surfaces. Quantum efficiency alone is not the same as overall energy conversion efficiency, as it does not convey information about the fraction of power that is converted by the solar cell.

Maximum Power Point

A solar cell may operate over a wide range of voltages (V) and currents (I). By increasing the resistive load on an irradiated cell continuously from zero (a *short circuit*) to a very high value (an *open circuit*) one can determine the maximum power point, the point that maximizes V×I; that is, the load for which the cell can deliver maximum electrical power at that level of irradiation. (The output power is zero in both the short circuit and open circuit extremes).

A high quality, monocrystalline silicon solar cell, at 25 °C cell temperature, may produce 0.60 V open-circuit (V_{OC}). The cell temperature in full sunlight, even with 25 °C air temperature, will probably be close to 45 °C, reducing the open-circuit voltage to 0.55 V per cell. The voltage drops modestly, with this type of cell, until the short-circuit current is approached (I_{SC}). Maximum power (with 45 °C cell temperature) is typically produced with 75% to 80% of the open-circuit voltage (0.43 V in this case) and 90% of the short-circuit current. This output can be up to 70% of the V_{OC} x I_{SC} product. The short-circuit current (I_{SC}) from a cell is nearly proportional to the illumination, while the open-circuit voltage (V_{OC}) may drop only 10% with an 80% drop in illumination. Lower-quality cells have a more rapid drop in voltage with increasing current and could produce only 1/2 V_{OC} at 1/2 I_{SC}. The usable power output could thus drop from 70% of the V_{OC} x I_{SC} product to 50% or even as little as 25%. Vendors who rate their solar cell "power" only as V_{OC} x I_{SC}, without giving load curves, can be seriously distorting their actual performance.

The maximum power point of a photovoltaic varies with incident illumination. For example, accumulation of dust on photovoltaic panels reduces the maximum power point. For systems large enough to justify the extra expense, a maximum power point tracker tracks the instantaneous power by continually measuring the voltage and current (and hence, power transfer), and uses this information to dynamically adjust the load so the maximum power is *always* transferred, regardless of the variation in lighting.

Fill Factor

Another defining term in the overall behavior of a solar cell is the fill factor (*FF*). This factor is a measure of quality of a solar cell. This is the available *power* at the maximum power point (P_m) divided by the open circuit voltage (V_{OC}) and the short circuit current (I_{SC}):

$$FF = \frac{P_m}{V_{OC} \times I_{SC}} = \frac{\eta \times A_c \times G}{V_{OC} \times I_{SC}}.$$

The fill factor can be represented graphically by the IV sweep, where it is the ratio of the different rectangular areas.

The fill factor is directly affected by the values of the cell's series, shunt resistances and diodes losses. Increasing the shunt resistance (R_{sh}) and decreasing the series resistance (R_s) lead to a higher fill factor, thus resulting in greater efficiency, and bringing the cell's output power closer to its theoretical maximum.

Typical fill factors range from 50% to 82%. The fill factor for a normal silicon PV cell is 80%.

Comparison

Energy conversion efficiency is measured by dividing the electrical output by the incident light power. Factors influencing output include spectral distribution, spatial distribution of power, temperature, and resistive load. IEC standard 61215 is used to compare the performance of cells and is designed around standard (terrestrial, temperate) temperature and conditions (STC): irradiance of 1 kW/m², a spectral distribution close to solar radiation through AM (airmass) of 1.5 and a cell temperature 25 °C. The resistive load is varied until the peak or maximum power point (MPP) is achieved. The power at this point is recorded as Watt-peak (Wp). The same standard is used for measuring the power and efficiency of PV modules.

Air mass affects output. In space, where there is no atmosphere, the spectrum of the sun is relatively unfiltered. However, on earth, air filters the incoming light, changing the solar spectrum. The filtering effect ranges from Air Mass 0 (AM0) in space, to approximately Air Mass 1.5 on Earth. Multiplying the spectral differences by the quantum efficiency of the solar cell in question yields the efficiency. Terrestrial efficiencies typically are greater than space efficiencies. For example, a silicon solar cell in space might have an efficiency of 14% at AM0, but 16% on earth at AM 1.5. However, that the number of incident photons in space is considerably larger, so the solar cell might produce considerably more power in space, despite the lower efficiency as indicated by reduced percentage of the total incident energy captured.

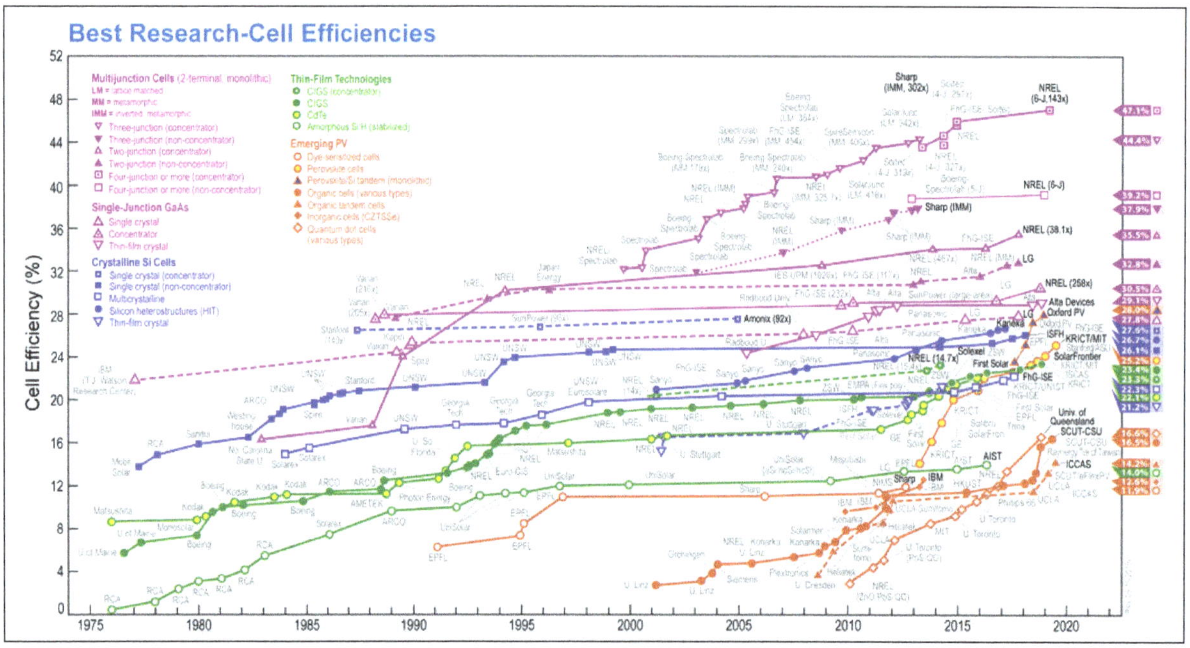

Reported timeline of research solar cell energy conversion efficiencies since 1976 (National Renewable Energy Laboratory).

Solar cell efficiencies vary from 6% for amorphous silicon-based solar cells to 44.0% with multiple-junction production cells and 44.4% with multiple dies assembled into a hybrid package. Solar cell energy conversion efficiencies for commercially available multicrystalline Si solar cells are around 14–19%. The highest efficiency cells have not always been the most economical — for example a 30% efficient multijunction cell based on exotic materials such as gallium arsenide or indium selenide produced at low volume might well cost one hundred times as much as an 8% efficient amorphous silicon cell in mass production, while delivering only about four times the output.

However, there is a way to "boost" solar power. By increasing the light intensity, typically photogenerated carriers are increased, increasing efficiency by up to 15%. These so-called "concentrator systems" have only begun to become cost-competitive as a result of the development of high efficiency GaAs cells. The increase in intensity is typically accomplished by using concentrating optics. A typical concentrator system may use a light intensity 6–400 times the sun, and increase the efficiency of a one sun GaAs cell from 31% at AM 1.5 to 35%.

A common method used to express economic costs is to calculate a price per delivered kilowatt-hour (kWh). The solar cell efficiency in combination with the available irradiation has a major influence on the costs, but generally speaking the overall system efficiency is important. Commercially available solar cells (as of 2006) reached system efficiencies between 5 and 19%.

Undoped crystalline silicon devices are approaching the theoretical limiting efficiency of 29.43%. In 2017, efficiency of 26.63% was achieved in an amorphous silicon/crystalline silicon heterojunction cell that place both positive and negative contacts on the back of the cell.

Energy Payback

The energy payback time is defined as the recovery time required for generating the energy spent for manufacturing a modern photovoltaic module. In 2008 it was estimated to be from 1 to 4 years depending on the module type and location. With a typical lifetime of 20 to 30 years, this means that modern solar cells would be net energy producers, i.e. they would generate more energy over their lifetime than the energy expended in producing them. Generally, thin-film technologies—despite having comparatively low conversion efficiencies—achieve significantly shorter energy payback times than conventional systems (often < 1 year).

A study published in 2013 which the existing literature found that energy payback time was between 0.75 and 3.5 years with thin film cells being at the lower end and multi-si-cells having a payback time of 1.5–2.6 years. A 2015 review assessed the energy payback time and EROI of solar photovoltaics. In this meta study, which uses an insolation of 1700 kWh/m^2/year and a system lifetime of 30 years, mean harmonized EROIs between 8.7 and 34.2 were found. Mean harmonized energy payback time varied from 1.0 to 4.1 years. Crystalline silicon devices achieve on average an energy payback period of 2 years.

Like any other technology, solar cell manufacture is dependent on and presupposes the existence of a complex global industrial manufacturing system. This comprises not only the fabrication systems typically accounted for in estimates of manufacturing energy, but the contingent mining, refining and global transportation systems, as well as other energy intensive critical support systems including finance, information, and security systems. The uncertainty of that energy component confers uncertainty on any estimate of payback times derived from that estimate, considered by some to be significant.

Technical Methods of Improving Efficiency

Choosing Optimum Transparent Conductor

The illuminated side of some types of solar cells, thin films, have a transparent conducting film to allow light to enter into the active material and to collect the generated charge carriers. Typically, films with high transmittance and high electrical conductance such as indium tin oxide, conducting polymers or conducting nanowire networks are used for the purpose. There is a trade-off between high transmittance and electrical conductance, thus optimum density of conducting nanowires or conducting network structure should be chosen for high efficiency.

Promoting Light Scattering in the Visible Spectrum

Lining the light-receiving surface of the cell with nano-sized metallic studs can substantially increase the cell efficiency of the cell. Light reflects off these studs at an oblique angle to the cell, increasing the length of the light path through the cell. This increases the number of photons absorbed by the cell and the amount of current generated.

The main materials used for the nano-studs are silver, gold, and aluminium. Gold and silver are not very efficient, as they absorb much of the light in the visible spectrum, which contains most of the energy present in sunlight, reducing the amount of light reaching the cell. Aluminium absorbs only ultraviolet radiation, and reflects both visible and infra-red light, so energy loss is minimized. Aluminium can increase cell efficiency up to 22% (in lab conditions).

Radiative Cooling

An increase in solar cell temperature of approximately 1 °C causes an efficiency decrease of about 0.45%. To prevent this, a transparent silica crystal layer can be applied to solar panels. The silica layer acts as a thermal black body which emits heat as infrared radiation into space, cooling the cell up to 13 °C.

Anti-reflective Coatings and Textures

Antireflective coatings could result in more destructive interference of incident light waves from the sun. Therefore, all sunlight would be transmitted into the photovoltaic. Texturizing, in which the surface of a solar cell is altered so that the reflected light strikes the surface again, is another technique used to reduce reflection. These surfaces can be created by etching or using lithography. Adding a flat back surface in addition to texturizing the front surface helps to trap the light within the cell, thus providing a longer optical path.

Rear Surface Passivation

Surface passivation is critical to solar cell efficiency. Many improvements have been made to the front side of mass-produced solar cells, but the aluminium back-surface is impeding efficiency improvements. The efficiency of many solar cells has benefitted by creating so-called passivated emitter and rear cells (PERCs). The chemical deposition of a rear-surface dielectric passivation layer stack that is also made of a thin silica or aluminium oxide film topped with a silicon nitride film helps to improve efficiency in silicon solar cells. This helped increase cell efficiency for commercial Cz-Si wafer material from just over 17% to over 21% by the mid-2010s, and the cell efficiency for quasi-mono-Si to a record 19.9%.

Concepts of the rear surface passivation for silicon solar cells has also been implemented for CIGS solar cells. The rear surface passivation shows the potential to improve the efficiency. Al2O3 and SiO2 have been used as the passivation materials. Nano-sized point contacts on Al2O3 layer and line contacts on SiO2 layer provide the electrical connection of CIGS absorber to the rear electrode Molybdenum. The point contacts on the Al2O3 layer are created by e-beam lithography and the line contacts on the SiO2 layer are created using photolithography. Also, the implementation of the passivation layers does not change the morphology of the CIGS layers.

Thin Film Materials

Thin film materials show a lot of promise for solar cells in terms of low costs and adaptability to existing structures and frameworks in technology. Since the materials are so thin, they lack the optical absorption of bulk material solar cells. Attempts to correct this have been tried, more important is thin film surface recombination. Since this is the dominant recombination process of nanoscale thin-film solar cells, it is crucial to their efficiency. Adding a passivating thin layer of silicon dioxide could reduce recombination.

NOMINAL POWER

The nominal power is the nameplate capacity of photovoltaic (PV) devices, such as solar cells, panels and systems, and is determined by measuring the electric current and voltage in a circuit, while varying the resistance under precisely defined conditions. These *Standard Test Conditions* (STC) are specified in standards such as IEC 61215, IEC 61646 and UL 1703; specifically the light intensity is 1000 W/m², with a spectrum similar to sunlight hitting the earth's surface at latitude 35°N in the summer (airmass 1.5), the temperature of the cells being 25 °C. The power is measured while varying the resistive load on the module between an open and closed circuit (between maximum and minimum resistance). The highest power thus measured is the 'nominal' power of the module in watts. This nominal power divided by the light power that falls on a given area of a photovoltaic device (area × 1000 W/m²) defines its efficiency, the ratio of the device's electrical output to the incident energy.

The nominal power is important for designing an installation in order to correctly dimension its cabling and converters. If the available area is limited the solar cell efficiency and with it the nominal power per area (e.g. kW/m²) is also relevant. For comparing modules, the price per nominal power (e.g. $/W) is relevant. For a given installation's physical orientation and location the expected annual production (e.g. kWh) per annual production assuming nominal power i.e. the capacity factor is important. With a projected capacity factor the price per projected annual production (e.g. $/kWh) can be estimated for a given installation. Finally, with a projected value of the production, the amortization of the cost of an installation can be estimated.

The peak power is not the same as the power under actual radiation conditions. In practice, this will be approximately 15-20% lower due to the considerable heating of the solar cells. Moreover, in installations where electricity is converted to AC, such as solar power plants, the actual total electricity generation capacity is limited by the inverter, which is usually sized at a lower peak capacity than the solar system for economic reasons. Since the peak DC power is reached only for a few hours each year, using a smaller inverter allows to save money on the inverter while clipping (wasting) only a very small portion of the total energy production. The capacity of the power plant after DC-AC conversion is usually reported in W_{AC} as opposed to W_p or W_{DC}.

Watt-peak

The International Bureau of Weights and Measures, which maintains the SI-standard, states that the physical unit and its symbol should not be used to provide specific information about a given physical quantity and that neither should be the sole source of information on a quantity. Nonetheless, colloquial English sometimes conflates the quantity power and its unit by using the non-SI unit watt-peak and the non-SI symbol W_p prefixed as within the SI, e.g. kilowatt-peak (kW_p), megawatt-peak (MW_p), etc. As such a photovoltaic installation may for example be described as having "one kilowatt-peak" in the meaning "one kilowatt of peak power". Similarly outside the SI, the peak power is sometimes written as "P = 1 kW_p" as opposed to "P_{peak} = 1 kW". In the context of domestic PV installations, the kilowatt (kW) is the most common unit for peak power, sometimes stated as kW_p.

Power Output in Real Conditions

The output of photovoltaic systems varies with the intensity of sunshine and other conditions. The more sun, the more power the PV module will generate. Losses, compared to performance in optimal conditions, will occur due to non-ideal alignment of the module in tilt and/or azimuth, higher temperature, module power mismatch (since panels in a system are connected in series the lowest performing module defines performance of the string it belongs to), soiling and DC to AC conversion. The power a module generates in real conditions can exceed the nominal power when the intensity of sunlight exceeds 1000 W/m^2 (which corresponds roughly to midday in summer in, for example, Germany), or when sun irradiation close to 1000 W/m^2 happens at lower temperatures.

Conversion from DC to AC

Solar power need to be converted from direct current (DC, as it is generated from the panel) to alternate current (AC) to be injected into the power grid. Since solar panels generate peak power only for few hours each year, and DC to AC converters are expensive, the converters are usually sized to be smaller than the peak DC power of the panels. This means that for few hours each year the peaks are "clipped" and the extra energy is lost. This has very little impact on the total energy generated throughout a year, but saves considerable amount of balance of system (BOS) costs. Due to under-sizing of converters, AC ratings of solar plants are generally significantly lower than DC ratings, as much as 30%. This in turn increases the calculated yearly capacity factor of the plant. The downrating of peak power and the related clipping is different from the losses incurred in the conversion from DC to AC, which happen at any power level and are usually relatively small.

Most countries refer to installed nominal nameplate capacity of PV systems and panels by counting DC power in watt-peak, denoted as W_p, or sometimes W_{DC}, as do most manufacturers and organizations of the photovoltaic industry, such as Solar Energy Industries Association (SEIA), the European Photovoltaic Industry Association (EPIA) or the International Energy Agency (IEA-PVPS). However, in some places of the world, a system's rated capacity is given after the power output has been converted to AC. These places include Canada, Japan (since 2012), Spain, and some parts of the United States. AC instead of DC is also given for most utility-scale PV power plants using CdTe-technology. Some grid regulations may limit the output of a PV system to as little as 70% of its nominal DC power (Germany). In such cases, the difference between nominal peak-power and converted AC output can therefore amount to as much as 30%. Because of these two different metrics, international organizations need to reconvert official domestic figures from the above-mentioned countries back to the raw DC output, in order to report coherent global PV-deployment in watt-peak.

In order to clarify whether the nominal power output (*"watt-peak"*, W_p) is in fact DC or already converted into AC, it is sometimes explicitly denoted as, for example, MW_{DC} and MW_{AC} or kW_{DC} and kW_{AC}. The converted W_{AC} is also often written as "MW (AC)", "MWac" or "MWAC". Just as for W_p, these units are non SI-compliant but widely used. In California, for example, where the rated capacity is given in MW_{AC}, a downrating of 15 percent in the conversion from DC to AC is assumed. This can be extremely confusing not only to non-experts, as the conversion efficiency has been improving to nearly 98 percent, grid regulations may change, some manufactures may differ from the rest of the industry, and countries, such as Japan, may adopt a different metric from one year to the other.

Cost-per-watt

Although watt-peak is a convenient measure, and is the standardized number in the photovoltaic industry on which prices, sales and growth numbers are based, it is arguably not the most important number for actual performance. Since a solar panel's job is to generate electric power at minimal cost, the amount of power that it generates under real-life conditions in relation to its cost should be the most important number to evaluate. This "cost-per-watt" measure is widely used in the industry.

It can happen that a panel from brand A and a panel of brand B give exactly the same watt-peak in laboratory test, but their power output is different in a real installation. This difference can be caused by different degradation rates at higher temperatures. At the same time, though brand A can be less productive than brand B it may as well cost less, thus it has a potential of becoming financially advantageous. An alternative scenario can also be true: a more expensive panel may produce so much more power that it will outperform a cheaper panel financially. An accurate analysis of long-term performance versus cost, both initial and on-going, is required to determine which panel may lead the owner to better financial results.

THERMOPHOTOVOLTAIC

Thermophotovoltaic (TPV) energy conversion is a direct conversion process from heat to electricity via photons. A basic thermophotovoltaic system consists of a thermal emitter and a photovoltaic diode cell.

The temperature of the thermal emitter varies between different systems from about 900 °C to about 1300 °C, although in principle TPV devices can extract energy from any emitter with temperature elevated above that of the photovoltaic device (forming an optical heat engine). The emitter can be a piece of solid material or a specially engineered structure. Thermal emission is the spontaneous emission of photons due to thermal motion of charges in the material. For these TPV temperatures, this radiation is mostly at near infrared and infrared frequencies. The photovoltaic diodes absorbs some of these radiated photons and converts them into electricity.

Thermophotovoltaic systems have few to no moving parts and are therefore quiet and require little maintenance. These properties make thermophotovoltaic systems suitable for remote-site and portable electricity-generating applications. Their efficiency-cost properties, however, are often poor compared to other electricity-generating technologies. Current research in the area aims at increasing system efficiencies while keeping the system cost low.

TPV systems usually attempt to match the optical properties of thermal emission (wavelength, polarization, direction) with the most efficient absorption characteristics of the photovoltaic cell, since unconverted thermal emission is a major source of inefficiency. Most groups focus on gallium antimonide (GaSb) cells. Germanium (Ge) is also suitable. Much research and development concerns methods for controlling the emitter's properties.

TPV cells have been proposed as auxiliary power conversion devices for capture of otherwise lost heat in other power generation systems, such as steam turbine systems or solar cells.

A prototype TPV hybrid car was built, the "Viking 29" (TPV) powered automobile, designed and built by the Vehicle Research Institute (VRI) at Western Washington University.

TPV research is an active area. Among others, the University of Houston TPV Radioisotope Power Conversion Technology development effort is attempting to combine a thermophotovoltaic cell with thermocouples to provide a 3 to 4-fold improvement in system efficiency over current radioisotope thermoelectric generators.

Thermophotovoltaics (TPVs) are a class of power generating system that convert thermal energy into electrical energy. They consist of, at a minimum, an emitter and a photovoltaic power converter. Most TPV systems include additional components such as concentrators, filters and reflectors.

The basic principle is similar to that of traditional photovoltaics (PV) where a p-n junction is used to absorb optical energy, generate and separate electron/hole pairs, and in doing so convert that energy into electricity. The difference is that the optical energy is not directly generated by the Sun, but instead by a material at high temperature (termed the emitter), that causes it to emit light. In this way thermal energy is converted to electrical energy.

The emitter can be heated by sunlight or other techniques. In this sense, TPVs provide a great deal of versatility in potential fuels. In the case of solar TPVs, large concentrators are needed to provide reasonable temperatures for efficient operation.

Improvements can take advantage of filters or selective emitters to create emissions in a wavelength range that is optimized for a specific photovoltaic (PV) converter. In this way TPVs can overcome a fundamental challenge for traditional PVs, making efficient use of the entire solar spectrum. For black body emitters, photons with energy less than the bandgap of the converter cannot be absorbed and are either reflected and lost or pass through the cell. Photons with energy above the bandgap can be absorbed, but the excess energy $\Delta G = E_{photon} - E_g$, is again lost, generating undesirable heating in the cell. In the case of TPVs, similar issues can exist, but the use of either selective emitters (emissivity over a specific wavelength range), or optical filters that only pass a narrow range of wavelengths and reflect all others, can be used to generate emission spectra that can be optimally converted by the PV device.

To maximize efficiency, all photons should be converted. A process often termed photon recycling can be used to approach this. Reflectors are placed behind the converter and anywhere else in the system that photons might not be efficiently directed to the collector. These photons are directed back to the concentrator where they can be converted, or back to the emitter, where they can be reabsorbed to generate heat and additional photons. An optimal TPV system would use photon recycling and selective emission to convert all photons into electricity.

Efficiency

The upper limit for efficiency in TPVs (and all systems that convert heat energy to work) is the Carnot efficiency, that of an ideal heat engine. This efficiency is given by:

$$\eta = 1 - \frac{T_{cell}}{T_{emit}}$$

where T_{cell} is the temperature of the PV converter. For the best reasonable values in a practical system, T_{cell}~300K and T_{emit}~1800, giving a maximum efficiency of ~83%. This limit sets the upper limit for the system efficiency. At 83% efficiency, all heat energy is converted to radiation by the emitter which is then converted by the PV into electrical energy without losses, such as thermalization or Joule heating. Maximum efficiency presumes no entropy change, which is only possible if the emitter and cell are at the same temperature. More accurate models are quite complicated.

Emitters

Deviations from perfect absorption and perfect black body behavior lead to light losses. For selective emitters, any light emitted at wavelengths not matched to the bandgap energy of the photovoltaic may not be efficiently converted and leads to reduced efficiency. In particular, emissions associated with phonon resonances are difficult to avoid for wavelengths in the deep infrared, which cannot be practically converted. Ideal emitters produce no infrared.

Filters

For black body emitters or imperfect selective emitters, filters reflect non-ideal wavelengths back to the emitter. These filters are imperfect. Any light that is absorbed or scattered and not redirected to the emitter or the converter is lost, generally as heat. Conversely, practical filters often reflect a small percentage of light in desired wavelength ranges. Both are inefficiencies.

Converters

Even for systems where only light of optimal wavelengths is passed to the converter, inefficiencies associated with non-radiative recombination and ohmic losses exist. Since these losses can depend on the light intensity incident on the cell, real systems must consider the intensity produced by a given set of conditions (emitter material, filter, operating temperature).

Geometry

In an ideal system, the emitter would be surrounded by converters so no light is lost. However, realistically, geometries must accommodate the input energy (fuel injection or input light) used to heat the emitter. Additionally, costs prohibit the placement of converters everywhere. When the emitter reemits light, anything that does not travel to the converters is lost. Mirrors can be used to redirect some of this light back to the emitter; however, the mirrors may have their own losses.

Black Body Radiation

For black body emitters where photon recirculation is achieved via filters, Planck's law states that a black body emits light with a spectrum given by:

$$I'(\lambda,T)=\frac{2hc^2}{\lambda^5}\frac{1}{e^{\frac{hc}{\lambda kT}}-1}$$

where I' is the flux of light of a specific wavelength, λ, given in units of $1/m^3/s$. h is Planck's constant, k is Boltzmann's constant, c is the speed of light, and T_{emit} is the emitter temperature. Thus, the light flux with wavelengths in a specific range can be found by integrating over the range. The peak wavelength is determined by the temperature, T_{emit} based on Wien's displacement law:

$$\lambda_{max}=\frac{b}{T}$$

where b is Wien's displacement constant. For most materials, the maximum temperature an emitter can stably operate at is about 1800 °C. This corresponds to an intensity that peaks at λ~1600 nm or an energy of ~0.75 eV. For more reasonable operating temperatures of 1200 °C, this drops to ~0.5 eV. These energies dictate the range of bandgaps that are needed for practical TPV converters (though the peak spectral power is slightly higher). Traditional PV materials such as Si (1.1 eV) and GaAs (1.4 eV) are substantially less practical for TPV systems, as the intensity of the black body spectrum is extremely low at these energies for emitters at realistic temperatures.

Active Components and Materials Selection

Emitters

Efficiency, temperature resistance and cost are the three major factors for choosing a TPV radiator. Efficiency is determined by energy absorbed relative to total incoming radiation. High temperature operation is a crucial factor because efficiency increases with operating temperature. As emitter temperature increases, black body radiation shifts to shorter wavelengths, allowing for more efficient absorption by photovoltaic cells. Cost is another major commercialization issue.

Polycrystalline Silicon Carbide

Polycrystalline silicon carbide (SiC) is the most commonly used emitter for burner TPVs. SiC is thermally stable to ~1700 °C. However, SiC radiates much of its energy in the long wavelength regime, far lower in energy than even the narrowest bandgap photovoltaic. This radiation is not converted into electrical energy. However, non-absorbing selective filters in front of the PV, or mirrors deposited on the back side of the PV can be used to reflect the long wavelengths back to the emitter, thereby recycling the unconverted energy. In addition, polycrystalline SiC is cheap to manufacture.

Tungsten

Refractory metals can be used as selective emitters for burner TPVs. Tungsten is the most common choice. It has higher emissivity in the visible and near-IR range of 0.45 to 0.47 and a low emissivity of 0.1 to 0.2 in the IR region. The emitter is usually in the shape of a cylinder with a sealed bottom, which can be considered a cavity. The emitter is attached to the back of a thermal absorber such as SiC and maintains the same temperature. Emission occurs in the visible and near IR range, which can be readily converted by the PV to electrical energy.

Rare-earth Oxides

Rare-earth oxides such as ytterbium oxide (Yb_2O_3) and erbium oxide (Er_2O_3) are the most commonly used selective emitters for TPVs. These oxides emit a narrow band of wavelengths in the near-infrared region, allowing the tailoring of the emission spectra to better fit the absorbance characteristics of a particular PV cell. The peak of the emission spectrum occurs at 1.29 eV for Yb_2O_3 and 0.827 eV for Er_2O_3. As a result, Yb_2O_3 can be used a selective emitter for Si PV cells and Er_2O_3, for GaSb or InGaAs. However, the slight mismatch between the emission peaks and band gap of the absorber results in a significant loss of efficiency. Selective emission only becomes significant at 1100 °C and increases with temperature, per Planck's Law. At operating temperatures below 1700 °C, selective emission of rare-earth oxides is fairly low, resulting in a further decrease in efficiency. Currently, 13% efficiency has been achieved with Yb_2O_3 and silicon PV cells. In general selective emitters have had limited success. More often filters are used with black body emitters to pass wavelengths matched to the bandgap of the PV and reflect mismatched wavelengths back to the emitter.

Photonic Crystals

Photonic crystals are a class of periodic materials that allow the precise control of electromagnetic wave properties. These materials give rise to the photonic bandgap (PBG). In the spectral range of the PBG, electromagnetic waves cannot propagate. The engineering of these materials allows some ability to tailor their emission and absorption properties, allowing for more effective design of selective emitters. Selective emitters with peaks at higher energy than the black body peak (for practical TPV temperatures) allow for wider bandgap converters. These converters are traditionally cheaper to manufacture and less temperature sensitive. Researchers at Sandia Labs demonstrated a high-efficiency (34% of light emitted from PBG selective emitter can be converted to electricity) TPV emitter using tungsten photonic crystals. However, manufacturing of these devices is difficult and not commercially feasible.

Photovoltaic Cells

Silicon

Early work in TPVs focused on the use of Si PVs. Silicon's commercial availability, extremely low cost, scalability and ease of manufacture makes this material an appealing candidate. However, the relatively wide bandgap of Si (1.1eV) is not ideal for use with a black body emitter at lower operating temperatures. Calculations using Planck's law, which describes the black body spectrum as a function of temperature, indicates that Si PVs would only be feasible at temperatures much higher than 2000 K. No emitter has been demonstrated that can operate at these temperatures. These engineering difficulties led to the pursuit of lower-bandgap semiconductor PVs.

Using selective radiators with Si PVs is still a possibility. Selective radiators would eliminate high and low energy photons, reducing heat generated. Ideally, selective radiators would emit no radiation beyond the band edge of the PV converter, increasing conversion efficiency significantly. No efficient TPVs have been realized using Si PVs.

Germanium

Early investigations into low bandgap semiconductors focused on germanium (Ge). Ge has a bandgap of 0.66 eV, allowing for conversion of a much higher fraction of incoming radiation. However, poor performance was observed due the extremely high effective electron mass of Ge. Compared to III-V semiconductors, Ge's high electron effective mass leads to a high density of states in the conduction band and therefore a high intrinsic carrier concentration. As a result, Ge diodes have fast decaying "dark" current and therefore, a low open-circuit voltage. In addition, surface passivation of germanium has proven extremely difficult.

Gallium Antimonide

The gallium antimonide (GaSb) PV cell, invented in 1989, is the basis of most PV cells in modern TPV systems. GaSb is a III-V semiconductor with the zinc blende crystal structure. The GaSb cell is a key development owing to its narrow bandgap of 0.72 eV. This allows GaSb to respond to light at longer wavelengths than silicon solar cell, enabling higher power densities in conjunction with manmade emission sources. A solar cell with 35% efficiency was demonstrated using a bilayer PV with GaAs and GaSb, setting the solar cell efficiency record.

Manufacturing a GaSb PV cell is quite simple. Czochralski Te-doped n-type GaSb wafers are commercially available. Vapor-based Zn diffusion is carried out at elevated temperatures ~450 °C to allow for p-type doping. Front and back electrical contacts are patterned using traditional photolithography techniques and an anti-reflective coating is deposited. Current efficiencies are estimated at ~20% using a 1000 °C black body spectrum. The radiative limit for efficiency of the GaSb cell in this setup is 52%, so vast improvements can still be made.

Indium Gallium Arsenide Antimonide

Indium gallium arsenide antimonide (InGaAsSb) is a compound III-V semiconductor. ($In_xGa_{1-x}As_ySb_{1-y}$) The addition of GaAs allows for a narrower bandgap (0.5 to 0.6 eV), and therefore better absorption of long wavelengths. Specifically, the bandgap was engineered to 0.55 eV. With this bandgap, the compound achieved a photon-weighted internal quantum efficiency of 79% with a fill factor of 65% for a black body at 1100 °C. This was for a device grown on a GaSb substrate by organometallic vapour phase epitaxy (OMVPE). Devices have been grown by molecular beam epitaxy (MBE) and liquid phase epitaxy (LPE). The internal quantum efficiencies (IQE) of these devices are approaching 90%, while devices grown by the other two techniques exceed 95%. The largest problem with InGaAsSb cells is phase separation. Compositional inconsistencies throughout the device degrade its performance. When phase separation can be avoided, the IQE and fill factor of InGaAsSb approach theoretical limits in wavelength ranges near the bandgap energy. However, the V_{oc}/E_g ratio is far from the ideal. Current methods to manufacture InGaAsSb PVs are expensive and not commercially viable.

Indium Gallium Arsenide

Indium gallium arsenide (InGaAs) is a compound III-V semiconductor. It can be applied in two ways for use in TPVs. When lattice-matched to an InP substrate, InGaAs has a bandgap of 0.74 eV, no better than GaSb. Devices of this configuration have been produced with a fill factor of 69%

and an efficiency of 15%. However, to absorb higher wavelength photons, the bandgap may be engineered by changing the ratio of In to Ga. The range of bandgaps for this system is from about 0.4 to 1.4 eV. However, these different structures cause strain with the InP substrate. This can be controlled with graded layers of InGaAs with different compositions. This was done to develop of device with a quantum efficiency of 68% and a fill factor of 68%, grown by MBE. This device had a bandgap of 0.55 eV, achieved in the compound $In_{0.68}Ga_{0.33}As$. n has the advantage of being a well-developed material. InGaAs can be made to lattice match perfectly with Ge resulting in low defect densities. Ge as a substrate is a significant advantage over more expensive or harder-to-produce substrates.

Indium Phosphide Arsenide Antimonide

The InPAsSb quaternary alloy has been grown by both OMVPE and LPE. When lattice-matched to InAs, it has a bandgap in the range 0.3–0.55 eV. The benefits of a TPV system with such a low band gap have not been studied in depth. Therefore, cells incorporating InPAsSb have not been optimized and do not yet have competitive performance. The longest spectral response from an InPAsSb cell studied was 4.3 µm with a maximum response at 3 µm. While this is a promising material, it has yet to be developed. For this and other low-bandgap materials, high IQE for long wavelengths is hard to achieve due to an increase in Auger recombination.

Lead Tin Selenide/Lead Strontium Selenide Quantum Wells

PbSnSe/PbSrSe quantum well materials, which can be grown by MBE on silicon substrates, have been proposed for low cost TPV device fabrication. These IV-VI semiconductor materials can have bandgaps between 0.3 and 0.6 eV. Their symmetric band structure and lack of valence band degeneracy result in low Auger recombination rates, typically more than an order of magnitude smaller than those of comparable bandgap III-V semiconductor materials.

Applications

TPVs promise efficient and economically viable power systems for both military and commercial applications. Compared to traditional nonrenewable energy sources, burner TPVs have little NO_x emissions and are virtually silent. Solar TPVs are a source of emission-free renewable energy. TPVs can be more efficient than PV systems owing to recycling of unabsorbed photons. However, TPVs are more complex and losses at each energy conversion step can lower efficiency. Further developments must be made to the absorber/emitter and PV cell. When TPVs are used with a burner source, they provide on-demand energy. As a result, energy storage is not needed. In addition, owing to the PV's proximity to the radiative source, TPVs can generate current densities 300 times that of conventional PVs.

Man-portable Power

Battlefield dynamics require portable power. Conventional diesel generators are too heavy for use in the field. Scalability allows TPVs to be smaller and lighter than conventional generators. Also, TPVs have few emissions and are silent. Multifuel operation is another potential benefit.

Early investigations into TPVs in the 1970s failed due to PV limitations. However, with the realization of the GaSb photocell, a renewed effort in the 1990s improved results. In early 2001, JX

Crystals delivered a TPV based battery charger to the Army that produced an output of 230 W with propane. This prototype utilized an SiC emitter operating at 1250 °C and GaSb photocells and was approximately 0.5 m tall. The power source had an efficiency of 2.5%, calculated by the ratio of the power generated to the thermal energy of the fuel burned. This is too low for practical battlefield use. To increase efficiency, narrow-band emitters must be realized and the temperature of the burner must be raised. Further thermal management steps, such as water cooling or coolant boiling, must be implemented. Although many successful proof-of-concept prototypes were demonstrated, no portable TPV power sources have reached troop testing or battlefield implementation.

Spacecraft

For space travel power generation systems must provide consistent and reliable power without large amounts of fuel. As a result, solar and radioisotope fuels (extremely high power density and long lifetime) are ideal sources of energy. TPVs have been proposed for each. In the case of solar energy, orbital spacecraft may be better locations for the large and potentially cumbersome concentrators required for practical TPVs. However, because of weight considerations and inefficiencies associated with the somewhat more complicated design of TPVs, conventional PVs will almost surely be more effective for these applications.

Probably more interesting is the prospect of using TPVs for conversion of radioisotope energy. The output of isotopes is thermal energy. In the past thermoelectricity (direct thermal to electrical conversion with no moving parts) has been used because of TPV efficiency is less than the ~10% of thermoelectric converters. Stirling engines have also been considered, but face reliability concerns, which are unacceptable for space missions, despite improved conversion efficiencies (>20%). However, with the recent advances in small-bandgap PVs, TPVs are becoming more promising candidates. A TPV radioisotope converter with 20% efficiency was demonstrated that uses a tungsten emitter heated to 1350 K, with tandem filters and a 0.6 eV bandgap InGaAs PV converter (cooled to room temperature). About 30% of the lost energy was due to the optical cavity and filters. The remainder was due to the efficiency of the PV converter.

Low-temperature operation of the converter is critical to the efficiency of TPV. Heating PV converters increases their dark current, thereby reducing efficiency. The converter is heated by the radiation from the emitter. In terrestrial systems it is reasonable to dissipate this heat without using additional energy with a heat sink. However, space is an isolated system, where heat sinks are impractical. Therefore, it is critical to develop innovative solutions to efficiently remove that heat, or optimized TPV cells that can operate efficiently with higher temperature converters. Both represent substantial challenges. Despite this, TPVs offer substantial promise for use in future space applications.

Commercial Applications

Off-grid Generators

Many homes are located in remote regions not connected to the power grid. Where available, power line extensions can be impractical. TPVs can provide a continuous supply of power in off-grid homes. Traditional PVs on the other hand, would not provide sufficient power during the winter months and nighttime, while TPVs can utilize alternative fuels to augment solar-only production.

The greatest advantage for TPV generators is cogeneration of heat and power. In cold climates, it can function as both a heater or stove and a power generator. JX Crystals developed a prototype TPV heating stove and generator. It burns natural gas and uses a SiC source emitter operating at 1250 °C and GaSb photocell to output 25,000 BTU/hr simultaneously generating 100 W. However, costs must be significantly reduced to render it commercially viable.

When a furnace is used as a heater and a generator, it is called combined heat and power (CHP). Many TPV CHP scenarios have been theorized, but a generator using boiling coolant was found most cost efficient. The proposed CHP would utilize a SiC IR emitter operating at 1425 °C and GaSb photocells cooled by boiling coolant. The TPV CHP would output 85,000 BTU/hr and generate 1.5 kW. The estimated efficiency would be 12.3% and the investment would be 0.08 €/kWh provided that the lifetime of the CHP furnace is 20 years. The estimated cost of other non-TPV CHPs are 0.12 €/kWh for gas engine CHP and 0.16 €/kWh for fuel cell CHP. This proposed furnace has not been commercialized because the market was not thought to be large enough.

Recreational Vehicles

TPVs have been proposed for use in recreational vehicles. With the advent of hybrid and other electrically powered vehicles, power generators with electrical outputs have become more interesting. In particular the versatility of TPVs for fuel choice and the ability to use multiple fuel sources makes them interesting as a wider variety of fuels are being with better sustainability are emerging. The silent operation of TPVs allow the generation of electricity when and where the use of noisy conventional generators is not allowed (i.e. during "quiet hours" in national park campgrounds), and do not disturb others. However, the emitter temperatures required for practical efficiencies make TPVs on this scale unlikely.

THERMODYNAMIC EFFICIENCY LIMIT

Thermodynamic efficiency limit is the absolute maximum theoretically possible conversion efficiency of sunlight to electricity. Its value is about 86%, which is the Chambadal-Novikov efficiency, an approximation related to the Carnot limit, based on the temperature of the photons emitted by the Sun's surface.

Effect of Band Gap Energy

Solar cells operate as quantum energy conversion devices, and are therefore subject to the thermodynamic efficiency limit. Photons with an energy below the band gap of the absorber material cannot generate an electron-hole pair, and so their energy is not converted to useful output and only generates heat if absorbed. For photons with an energy above the band gap energy, only a fraction of the energy above the band gap can be converted to useful output. When a photon of greater energy is absorbed, the excess energy above the band gap is converted to kinetic energy of the carrier recombination. The excess kinetic energy is converted to heat through phonon interactions as the kinetic energy of the carriers slows to equilibrium velocity. Hence, the solar energy cannot be converted to electricity beyond a certain limit.

Solar cells with multiple band gap absorber materials improve efficiency by dividing the solar spectrum into smaller bins where the thermodynamic efficiency limit is higher for each bin. The thermodynamic limits of such cells (also called multi-junction cells, or tandem cells) can be analyzed using and online simulator in nanoHUB.

Efficiency Limits for Different Solar Cell Technologies

Thermodynamic efficiency limits for different solar cell technologies are as follows:

- Single junctions ≈ 33%
- 3-cell stacks and impure PVs ≈ 50%
- Hot carrier- or impact ionization-based devices ≈ 54-68%
- Commercial modules are ≈ 12-21%
- Solar cell with an upconverter for operation in the AM1.5 spectrum and with a 2eV bandgap ≈ 50.7%

Thermodynamic Efficiency Limit for Excitonic Solar Cells

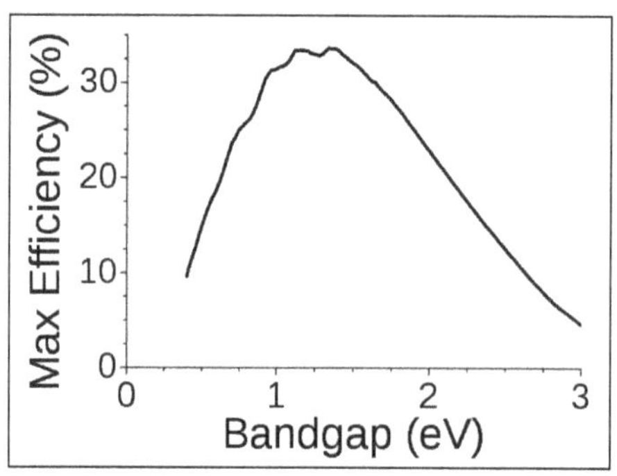

The Shockley-Queisser limit for the efficiency of a single-junction solar cell under unconcentrated sunlight. This calculated curve uses actual solar spectrum data, and therefore the curve is wiggly from IR absorption bands in the atmosphere. This efficiency limit of about 34% can be exceeded by multijunction solar cells.

Excitonic solar cells generates free charge by bound and intermediate exciton states unlike inorganic and crystalline solar cells. The efficiency of the excitonic solar cells and inorganic solar cells (with less exciton-binding energy) cannot go beyond 31% as explained by Shockley and Queisser.

Thermodynamic Efficiency Limits with Carrier Multiplication

Carrier multiplication facilitates multiple electron-hole pair generation for each photon absorbed. Efficiency limits for photovoltaic cells can be theoretically higher considering thermodynamic effects. For a solar cell powered by the Sun's unconcentrated black-body radiation, the theoretical maximum efficiency is 43% whereas for a solar cell powered by the Sun's full concentrated radiation, the efficiency limit is up to 85%. These high values of efficiencies are possible only when the solar cells use radiative recombination and carrier multiplication.

SUN-FREE PHOTOVOLTAICS

Sun-free photovoltaics is a photovoltaics technology which does not require sunlight to produce electricity. This technique was developed by research team at Massachusetts Institute of Technology Photovoltaic cells convert light to electricity most efficiently at specific wavelengths. The surface featured of Sun-free photovoltaics is engineered such that it converts heat energy into the specific wavelengths. This increases the efficiency of existing thermophotovoltaic (TPV) systems.

Principle of Working

The surface of this material is etched with billions of nanoscale pits, so that it emits wave at the specified wavelengths and suppresses other wavelengths which are not required. When the surface is exposed to heat energy, it radiates light energy at required wavelengths. The atoms liberates free electrons when radiation is applied, thus generating electricity.

Development

A device that uses radioisotopes is capable of producing electricity for up to 30 years without interruption using a process called radioactive decay. Another similar device was developed by researchers from MIT, which uses butane as a fuel source. It has the size of a button and is estimated to operate three times longer than a similarly sized lithium-ion battery and has the advantage of recharging immediately upon refuelling. Thermo-photovoltaics, which is a technology developed in the 1960s, is capable of converting heat energy into electricity. It was previously known that sunlight is not absolutely necessary for photovoltaic cells to operate. Infrared radiation is not utilized efficiently in photovoltaic materials which are characterized by low band gap, though the performance is comparatively better than normal silicon photovoltaic cells. One example of photovoltaics utilizing heat is a photovoltaic diode which produces electricity from heat given off by a hydrocarbon fuelled thermal emitter.

In Sun-free photovoltaics, the thermal emitter described above is finely tuned to emit specific wavelengths that the photovoltaic diode can utilize, while at the same time stopping the wavelengths that cannot be used by the diode. This is achieved by etching features of nanoscale like ridges and holes on a photonic crystal so that the light passing through the crystal can be modified.

Operation

A slab of tungsten is used as a thermal emitter with billions of nanoscale pits on the surface, with each pit acting like a resonator. On heating, specific wavelengths are radiated by the slab as it generates an altered emission spectrum.

Applications

Sun-free photovoltaics can be applied in electronic devices like smartphones. Its value of current density has the capability to be tripled in the coming years, when it could power a smartphone for more than 20 days without needing to recharge in between. It can also be used to utilize waste heat from electric appliances like televisions and electronic appliances like mobile phones.

POLARIZING ORGANIC PHOTOVOLTAICS

Polarizing organic photovoltaics (ZOPV) is a concept for harvesting energy from Liquid crystal display screens, developed by engineers from UCLA. This concept enables devices to utilize external light and the LCD screen's backlight using photovoltaic polarizers. Photovoltaic polarizers convert this light into electricity which can be used to power the device. This concept also provides multifunctional capability to devices with LCD screens as they act as photovoltaic devices and also as polarisers.

A liquid crystal display (LCD) is a flat panel display, electronic visual display, video display that uses the light modulating properties of liquid crystals (LCs). LCs do not emit light directly. They are used in a wide range of applications, including computer monitors, television, instrument panels, laptops tablet computers etc. They are common in consumer devices such as video players, gaming devices, clocks, watches, calculators, and telephones.

Operation

Up to three-fourths of the light energy wasted from LCD backlight illumination can be retrieved and utilized using polarizing organic photovoltaics. They can utilize external light energy also apart from backlight illumination using photovoltaic polarizers, which are present within the structure of the LCD screen.

Advantages

80% to 90% of the total energy utilized by any device with an LCD screen is used up by the backlight illumination. As polarizing organic photovoltaics can recycle up to 75% of wasted light energy, the efficiency of the device is increased.

Disadvantages

This simply incorporates additional conversion efficiency losses. These devices harvest their own light.

CADMIUM TELLURIDE PHOTOVOLTAICS

Cadmium telluride (CdTe) photovoltaics describes a photovoltaic (PV) technology that is based on the use of cadmium telluride, a thin semiconductor layer designed to absorb and convert sunlight into electricity. Cadmium telluride PV is the only thin film technology with lower costs than conventional solar cells made of crystalline silicon in multi-kilowatt systems.

On a lifecycle basis, CdTe PV has the smallest carbon footprint, lowest water use and shortest energy payback time of any current photo voltaic technology. CdTe's energy payback time of less than a year allows for faster carbon reductions without short-term energy deficits.

The toxicity of cadmium is an environmental concern mitigated by the recycling of CdTe modules at the end of their life time, though there are still uncertainties and the public opinion is skeptical towards this technology. The usage of rare materials may also become a limiting factor to the industrial scalability of CdTe technology in the mid-term future. The abundance of tellurium—of which telluride is the anionic form—is comparable to that of platinum in the earth's crust and contributes significantly to the module's cost.

PV array made of cadmium telluride (CdTe) solar panels.

CdTe photovoltaics are used in some of the world's largest photovoltaic power stations, such as the Topaz Solar Farm. With a share of 5.1% of worldwide PV production, CdTe technology accounted for more than half of the thin film market in 2013. A prominent manufacturer of CdTe thin film technology is the company First Solar, based in Tempe, Arizona.

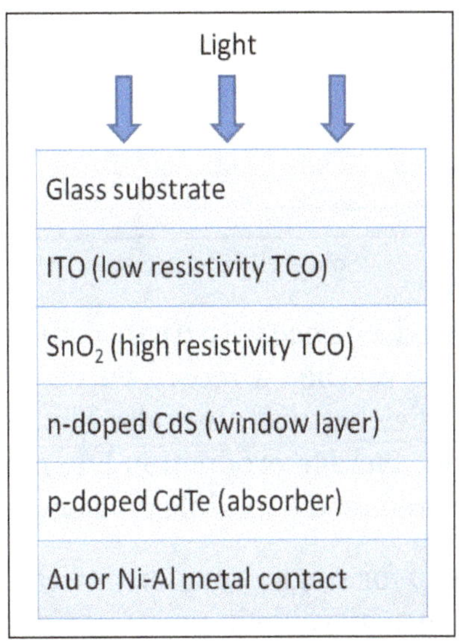

Cross-section of a CdTe thin film solar cell.

The dominant PV technology has always been based on crystalline silicon wafers. Thin films and concentrators were early attempts to lower costs. Thin films are based on using thinner semiconductor layers to absorb and convert sunlight. Concentrators lower the number of panels by using lenses or mirrors to put more sunlight on each panel.

The first thin film technology to be extensively developed was amorphous silicon. However, this technology suffers from low efficiencies and slow deposition rates (leading to high capital costs).

Instead, the PV market reached some 4 gigawatts in 2007 with crystalline silicon comprising almost 90% of sales. The same source estimated that about 3 gigawatts were installed in 2007.

During this period cadmium telluride and copper indium diselenide or CIS-alloys remained under development. The latter is beginning to be produced in volumes of 1–30 megawatts per year due to very high, small-area cell efficiencies approaching 20% in the laboratory. CdTe cell efficiency is approaching 20% in the laboratory with a record of 22.1% as of 2016.

Technology

Cell Efficiency

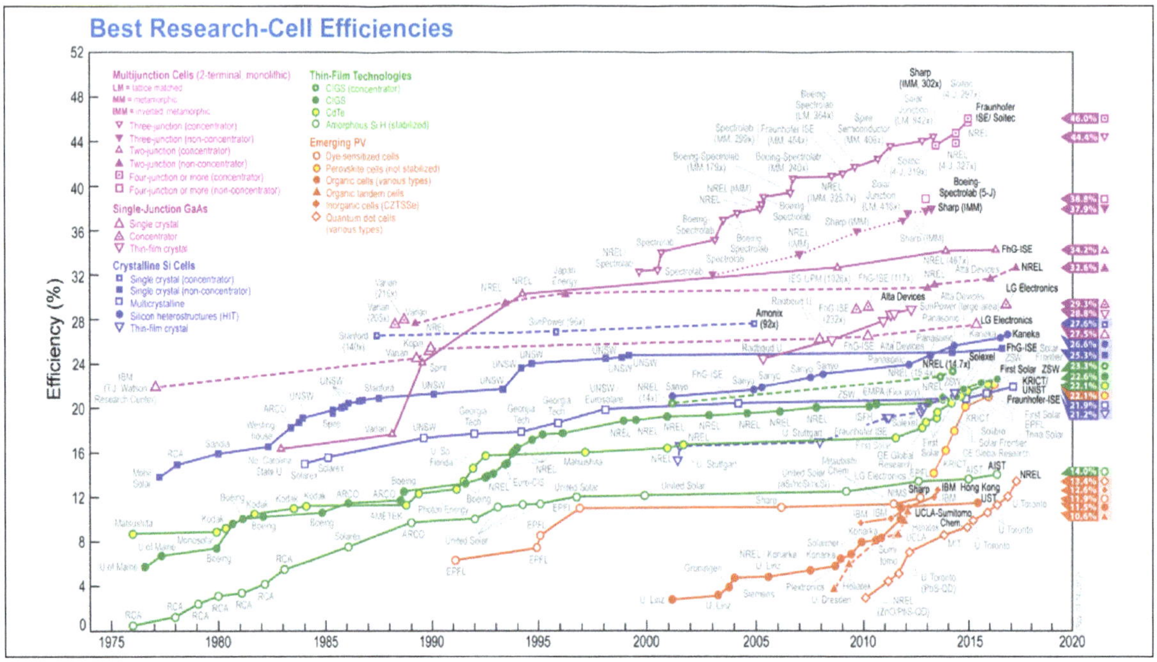

Solar cell efficiencies.

In August 2014 First Solar announced a device with 21.1% conversion efficiency. In February 2016, First Solar announced that they had reached a record 22.1% conversion efficiency in their CdTe cells. In 2014, the record module efficiency was also raised by First Solar from 16.1% up to 17.0%. At this time, the company projected average production line module efficiency for its CdTe PV to be 17% by 2017, but by 2016, they predicted a module efficiency closer to ~19.5%.

Since CdTe has the optimal band gap for single-junction devices, efficiencies close to 20% (such as already shown in CIS alloys) may be achievable in practical CdTe cells.

Process Optimization

Process optimization improved throughput and lowered costs. Improvements included broader substrates (since capital costs scale sublinearly and installation costs can be reduced), thinner layers (to save material, electricity, and processing time), and better material utilization (to save material and cleaning costs). 2014 CdTe module costs were about $72 per 1 square metre (11 sq ft), or about $90 per module.

Ambient Temperature

Module efficiencies are measured in laboratories at standard testing temperatures of 25 °C, however in the field modules are often exposed to much higher temperatures. CdTe's relatively low temperature coefficient protects performance at higher temperatures. CdTe PV modules experience half the reduction of crystalline silicon modules, resulting in an increased annual energy output of 5-9%.

Solar Tracking

Almost all thin film photovoltaic module systems to-date have been non-solar tracking, because module output was too low to offset tracker capital and operating costs. But relatively inexpensive single-axis tracking systems can add 25% output per installed watt. In addition, depending on the Tracker Energy Gain, the overall eco-efficiency of the PV system can be enhanced by lowering both system costs and environmental impacts. This is climate-dependent. Tracking also produces a smoother output plateau around midday, better matching afternoon peaks.

Materials

Cadmium

Cadmium (Cd), a toxic heavy metal considered a hazardous substance, is a waste byproduct of mining, smelting and refining sulfidic ores of zinc during zinc refining, and therefore its production does not depend on PV market demand. CdTe PV modules provide a beneficial and safe use for cadmium that would otherwise be stored for future use or disposed of in landfills as hazardous waste. Mining byproducts can be converted into a stable CdTe compound and safely encapsulated inside CdTe PV solar modules for years. A large growth in the CdTe PV sector has the potential to reduce global cadmium emissions by displacing coal and oil power generation.

Tellurium

Tellurium (Te) production and reserves estimates are subject to uncertainty and vary considerably. Tellurium is a rare, mildly toxic metalloid that is primarily used as a machining additive to steel. Te is almost exclusively obtained as a by-product of copper refining, with smaller amounts from lead and gold production. Only a small amount, estimated to be about 800 metric tons per year, is available. According to USGS, global production in 2007 was 135 metric tons. One gigawatt (GW) of CdTe PV modules would require about 93 metric tons (at current efficiencies and thicknesses). Through improved material efficiency and increased PV recycling, the CdTe PV industry has the potential to fully rely on tellurium from recycled end-of-life modules by 2038. In the last decade, new supplies have been located, e.g., in Xinju, China as well as in Mexico and Sweden. In 1984 astrophysicists identified tellurium as the universe's most abundant element having an atomic number over 40. Certain undersea ridges are rich in tellurium.

Cadmium Chloride/Magnesium Chloride

The manufacture of a CdTe cell includes a thin coating with cadmium chloride ($CdCl_2$) to increase the cell's overall efficiency. Cadmium chloride is toxic, relatively expensive and highly soluble in water, posing a potential environmental threat during manufacture. In 2014 research discovered that abundant and harmless magnesium chloride ($MgCl_2$) performs as well as cadmium chloride. This research may lead to cheaper and safer CdTe cells.

Safety

By themselves, cadmium and tellurium are toxic and carcinogenic, but CdTe forms a crystalline lattice that is highly stable, and is several orders of magnitude less toxic than cadmium. The glass plates surrounding CdTe material sandwiched between them (as in all commercial modules) seal during a fire and do not allow any cadmium release unless the glass is broken. All other uses and exposures related to cadmium are minor and similar in kind and magnitude to exposures from other materials in the broader PV value chain, e.g., to toxic gases, lead solder, or solvents (most of which are not used in CdTe manufacturing).

Recycling

Due to the exponential growth of photovoltaics the number of worldwide installed PV systems has increased significantly. First Solar established the first global and comprehensive recycling program in the PV industry in 2005. Its recycling facilities operate at each of First Solar's manufacturing plants and recover up to 95% of semiconductor material for reuse in new modules and 90% of glass for reuse in new glass products. A life cycle assessment of CdTe module recycling by the University of Stuttgart showed a reduction in primary energy demand in End-Of-Life from 81 MJ/m^2 to -12 MJ/m^2, a reduction of around 93 MJ/m^2, and in terms of global warming potential from 6 kg CO2-equiv./m^2 to -2.5 CO2-equiv./m^2, a reduction of around -8.5 CO2-equiv./m^2. These reductions show a highly beneficial change in the overall environmental profile of CdTe photovoltaic module. The LCA also showed that the main contributors to considered environmental impact categories are due to required chemicals and energy within the processing of CdTe modules.

Grain Boundaries

Grain boundary is the interface between two grains of a crystalline material and occur when two grains meet. They are a type of crystalline defect. It is often assumed that the open-circuit voltage gap seen in CdTe, in comparison to both single crystal GaAs and the theoretical limit, may be in some way attributable to the grain boundaries within the material. There have however been a number of studies which have suggested not only that GBs are not deleterious to performance but may in fact be beneficial as sources of enhanced carrier collection. So, the exact role of the grain boundaries in limitation of performance of CdTe-based solar cells remains unclear and the research is ongoing to address this question.

MAXIMUM POWER POINT TRACKING

Maximum Power Point Tracking (MPPT) is a technology approach used in solar PV inverters to optimise power output in less-than-ideal sunlight conditions. Most modern inverters are equipped with at least one MPPT input.

An MPPT tracker is analogous to a thumb placed over a garden hose. If you put your thumb over part of the opening of the hose (adding resistance to the circuit), the pressure (voltage) goes up and the stream flies faster, but less water (current) is getting through. If you completely cover the opening, nothing gets through. If you remove your thumb entirely, the maximum flowrate gets through, but the stream falls limply at your feet. That's the basic mechanism of the MPPT tracker: vary resistance in the circuit to modify current and voltage.Now imagine that there are hundreds of pumps (solar panels) upstream of the hose and they are delivering water (energy) to you. Further complicating things, some of these pumps go offline at certain parts of the day (partial shading of the array). So the force behind the delivery of water will be constantly varying.

Your task is to wash your car, which is sitting still about 15 feet away. You need to keep moving your thumb as the upstream pump force varies in order to avoid undershooting or overshooting the car. The "car" in this case is the Maximum Power Point – for any array of solar panels, there is a configuration of current and voltage that aligns with maximum power generation:

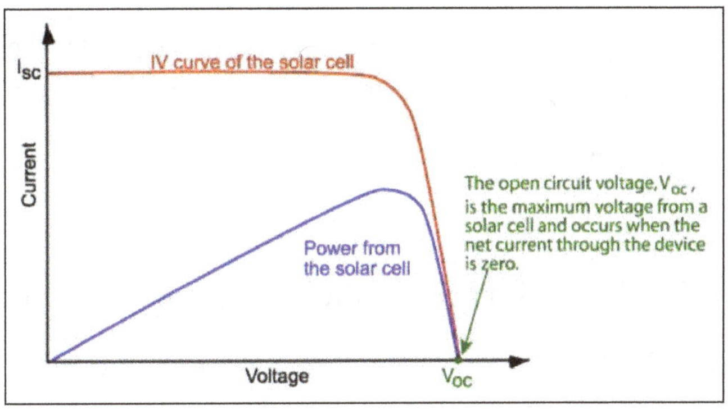

The MPPT tracker varies resistance in order to keep hitting this point, using control logic to stay at the maximum just like a thermostat or cruise control.

Block Diagram of MPPT

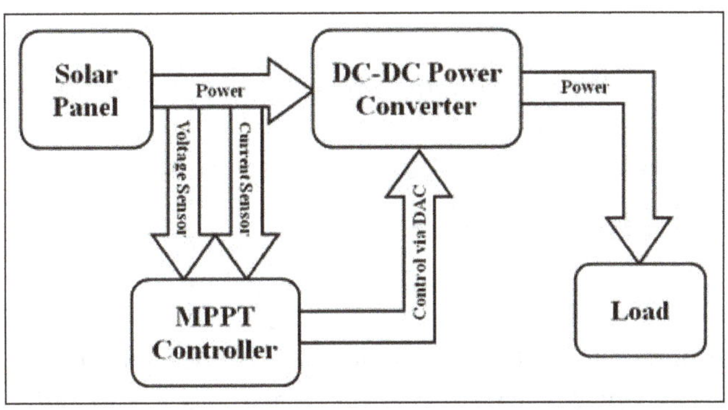

Block diagram of MPPT algorithm.

The maximum power tracking mechanism makes use of an algorithm and an electronic circuitry. The mechanism is based on the principle of impedance matching between load and PV module, which is necessary for maximum power transfer. Generally MPPT is an adaptation of DC to DC switching voltage regulator. The impedance matching is done by using a DC to DC converter. Using a DC to DC converter the impedance is matched by changing the duty cycle of the switch. Coupling to the load for maximum power transfer may require either providing a higher voltage for higher current. A buck boost scheme is commonly used with voltage and current sensors tied into a feedback loop using a controller to vary the switching times. Buck converter can also be used. The block diagram of an MPPT algorithm is shown in above figure.

The power from the solar module can be calculated by measuring the voltage and the current. This power is an input to the algorithm which then adjusts the duty cycle of the switch, resulting in the adjustment of the reflected load impedance according to the power output of PV module. For instance the relation between the input voltage (Vi) and the output voltage (Vo) and the impedance of load (R_L) reflected at the input side (Ri) of a buck type DC to DC converter can be given as,

$$Vo = Vi \times d$$

$$Ri = R_L / d^2$$

Where d is the duty cycle. By adjusting the duty cycle, Ri can be varied which should be same as the impedance of solar PV module (R_{PV}) in a given operating condition for maximum power transfer.

The power output of a PV system is given by,

$$P = V \times I$$

With incremental change in current and voltage, the modified power is given by,

$$P + \delta P = (I + \delta I) \times (V + \delta V)$$

This after ignoring small terms simplifies to,

$$\delta P = (\delta V \times I) + (\delta I \times V)$$

δP must be zero at break point. Therefore at peak point the above expression in the limit becomes,

$$dV / dI = -V / I$$

It may be noted here that dV / dI is the dynamic impedance of the source, which is required to be equal to negative of static impedance, V / I .

Strategies for Operation of an MPPT

There are three possible strategies for the operation of an MPPT. They are:

- By monitoring dynamic and static impedances: A small signal current is periodically injected into the array bus and the dynamic as well as static bus impedances (Zd and Zs respectively) are measured. The operating voltage is then adjusted until the condition Zd = - Zs is achieved.
- By monitoring power output: From the shape of a PV characteristic, it is clear that the slope, dP / dV is zero at maximum power point. This property is utilized to track the minimum power point. Voltage is adjusted and power output is sensed. The operating voltage is increased as long as dP / dV is positive. That is, voltage is increased as long as we get

increased output. If dP / dV is sensed negative, the operating voltage is decreased. The voltage is held unaltered if dP /dV is near zero within a preset dead band.

- By fixing the output voltage as a fraction of Voc: This method makes use of the fact that for most PV cells the ratio of the voltage at maximum power point to open circuit voltage, is approximately constant (say k). For high quality crystalline silicon cell, k=0.72.

Algorithms for MPPT

Algorithms for MPPT are various types of schemes that are implemented for obtaining maximum power transfer. Some of the popular schemes are incremental conductance method, system oscillation method, hill climbing method, modified hill climbing method, constant voltage method. Other MPPT methods include those which use state space approach with the tracking power converter operating in Continuous Conduction Mode (CCM) and the another one which is based on a combination of incremental conductance and perturb and observe method. Energy extracted from the PV source through MPPT should be either utilized by a load or stored in some form for example, energy stored in a battery or used for electrolysis to produce hydrogen for future use in fuel cells. In view of this grid connected PV systems are very popular as they do not have any energy storage requirements since the grid can absorb any amount of PV energy tracked.

Some of the popular and most commonly used MPPT schemes are explained below:

Constant voltage method: The ration of V_{MPP} and Voc is a constant approxiately equal to 0.78. Here array voltage is represented by V_{MPP} and the open circuit voltage is represented by Voc. The sensed PV array voltage is compared with a reference voltage to generate an error signal which in turn controls the duty cycle. The duty cycle of the power converter ensures that the PV array voltage is equal to 0.78 × Voc. Also Voc can be determined using a diode mounted at the back of the array (so that it has the same temperature as the array). A constant current is fed into the diode and the resulting voltage across the diode is used as the arrays Voc which then utilized in tracking V_{MPP}.

Hill climbing method: The most popular algorithm is the hill climbing method. It is applied by perturbing the duty cycle 'd' at regular intervals and by recording the resulting array current and voltage values, thereby obtaining the power. Once the power is known, a check for the slope of the P- V curve or the operating region (current source or voltage source region) is carried out and then the change in d is effected in a direction so that the operating point approaches maximum power point on the power voltage characteristic. The algorithm of this scheme is described below along with the help of mathematical expressions,

In a voltage source region, $\partial P_{PV} / \partial V_{PV} > 0 = d = d + \delta d$ (i.e., increment d)

In the current source region, $\partial P_{PV} / \partial V_{PV} < 0 = d = d - \delta d$ (i.e., decrement d)

At maximum power point, $\partial P_{PV} / \partial V_{PV} = 0 = d = d$ or $\delta d = 0$ (i.e., retain d)

As per the above equations if $\partial P_{PV} / \partial V_{PV}$ is graeter than zero ($P_{new} > P_{old}$) the duty cycle is increased ($d = d + \delta d$). This means that the slope is positive and the module is operating in the constant current region. In case of the slope being negative ($P_{new} < P_{old}$) the duty cycle is reduced ($d = d - \delta d$), as the operating region in this case is the constsnt voltage region. This algorithm can be implemented using a microcontroller.

Incremental conductance method: In the Incremental conductance method, the maximum power point by matching the PV array impedance wit the effective impedance of the converter reflected across the terminals of the array. While, latter is tuned by increase or decrease in the duty cycle value. The algorithm can be explained as follows:

For voltage source region, $\partial I_{PV} / \partial V_{PV} > - I_{PV} / V_{PV} = d = d + \delta d$ (i.e., increment duty cycle)

For current source region, $\partial I_{PV} / \partial V_{PV} < - I_{PV} / V_{PV} = d = d - \delta d$ (i.e., decrement duty cycle)

At maximum power point, $\partial I_{PV} / \partial V_{PV} = d = d$ or $\delta d = 0$

Incremental conductance MPPT method: Off-grid PV system are usually using batteries to supply loads at night. Although the fully charged battery pack voltage may be near to the maximum power point voltage of the PV panel's, this is not true at sunrise when the partial discharge of the battery takes place. At a certain voltage below the maximum voltage of the PV panel, charging takes place and this mismatch can be resolved using an MPPT. In case of a grid connected PV system, all the delivered power from solar modules will be sent to the grid. Therefore, the MPPT in a grid connected photovoltaic system will always try to operate the PV modules at its maximum power point.

Advantages of MPPT

The advantages of MPPT depend on your array, your climate, and your seasonal load pattern. It gives you an effective current boost only when the Vpp is more than about 1V higher than the battery voltage. In hot weather, this may not be the case unless the batteries are low in charge. In cold weather however, the Vpp can rise to 18V. If your energy use is greatest in the winter (typical in most homes) and you have cold winter weather, then you can gain a substantial boost in energy when you need it the most!

Example: MPPT on a Cold Winter Day

If the outside temperature is 20 °F (-7 °C) and the wind is blowing a bit, the PV cell temperature rises to only around 32 °F (0 °C). The Vpp = 18V. Batteries are a bit low, and loads are on, so the battery voltage = 12.0.

The ratio of Vpp to battery voltage is 18:12 = 1.5:1.

Under these conditions, a theoretically perfect MPPT (with no voltage drop in the array circuit) would deliver a 50% increase in charge current. In reality, there are losses in the conversion just as there is friction in a car's transmission. Reports from the field indicate that increases of 20 to 30% are typically observed.

References

- Nelson, R.E. (2003). "A brief history of thermophotovoltaic development". Semiconductor Science and Technology. 18 (5): S141–S143. Bibcode:2003sesct..18S.141N. Doi:10.1088/0268-1242/18/5/301
- Solar+irradiance: encyclopedia.thefreedictionary.com, Retrieved 26 July, 2019
- Steinhilber, F.; Beer, J.; Fröhlich, C. (2009). "Total solar irradiance during the Holocene". Geophys. Res. Lett. 36 (19): L19704. Bibcode:2009georl..3619704S. Doi:10.1029/2009GL040142

- Solar-radiation-basics, solar: energy.gov, Retrieved 21 May, 2019
- Stickler, Greg. "Educational Brief - Solar Radiation and the Earth System". National Aeronautics and Space Administration. Archived from the original on 25 April 2016. Retrieved 5 May 2016
- B Kumar, Ankush (3 January 2017). "Predicting efficiency of solar cells based on transparent conducting electrodes". Journal of Applied Physics. 121 (1): 014502. Bibcode:2017JAP...121a4502k. Doi:10.1063/1.4973117. ISSN 0021-8979
- Sun-Free Photovoltaics: Materials Engineered to Give Off Precisely Tuned Wavelengths of Light When Heated". Sciencedaily.com. Retrieved 2011-08-08
- How-does-maximum-power-tracking-work: solarchoice.net.au, Retrieved 8 January, 2019

Photovoltaic Systems

CHAPTER 3

Photovoltaic systems are the power systems that are designed to supply solar power by means of photovoltaics. The components of photovoltaic systems include solar cell, solar panel, solar tracker and solar inverter. The topics elaborated in this chapter will help in gaining a better perspective about photovoltaic systems and its components.

SOLAR CELL

A solar cell or photovoltaic cell is a device that converts solar energy into electricity by the photovoltaic effect. Sometimes, the term solar cell is reserved for devices intended specifically to capture energy from sunlight, while the term photovoltaic cell is used when the source is unspecified. Assemblies of cells are used to make solar panel, solar modules, or photovoltaic arrays. Photovoltaics is the field of technology and research related to the application of solar cells for solar energy.

Three Generations of Solar Cells

Solar Cells are classified into three generations which indicates the order of which each became important. At present there is concurrent research into all three generations while the first generation technologies are most highly represented in commercial production, accounting for 89.6 percent of 2007 production.

First Generation

First generation cells consist of large-area, high quality and single junction devices. First Generation technologies involve high energy and labor inputs which prevent any significant progress in reducing production costs. Single junction silicon devices are approaching the theoretical limiting efficiency of 33 percent and achieve cost parity with fossil fuel energy generation after a payback period of 5-7 years. They are not likely to get lower than US$1/W.

Second Generation

Second generation materials have been developed to address energy requirements and production costs of solar cells. Alternative manufacturing techniques such as vapor deposition, electroplating, and use of Ultrasonic Nozzles are advantageous as they reduce high temperature processing significantly. It is commonly accepted that as manufacturing techniques evolve production costs will be dominated by constituent material requirements, whether this be a silicon substrate, or glass cover. Second generation technologies are expected to gain market share in 2008.

Such processes can bring costs down to a little under US$0.50/W but because of the defects inherent in the lower quality processing methods, have much reduced efficiencies compared to First Generation.

The most successful second generation materials have been cadmium telluride (CdTe), copper indium gallium selenide, amorphous silicon and micromorphous silicon. These materials are applied in a thin film to a supporting substrate such as glass or ceramics reducing material mass and therefore costs. These technologies do hold promise of higher conversion efficiencies, particularly CIGS-CIS, DSC and CdTe offers significantly cheaper production costs.

Among major manufacturers there is certainly a trend toward second generation technologies however commercialization of these technologies has proven difficult. In 2007 First Solar produced 200 MW of CdTe solar cells making it the fifth largest producer of solar cells in 2007 and the first ever to reach the top ten from production of second generation technologies alone. Honda Soltec Co., Ltd., a Honda company, also began to commercialize] their CIGS base solar panel in 2008.

In 2007 CdTe production represented 4.7 percent of total market share, thin-film silicon 5.2 percent and CIGS 0.5 percent.

Third Generation

Third generation technologies aim to enhance poor electrical performance of second generation (thin-film technologies) while maintaining very low production costs.

Current research is targeting conversion efficiencies of 30-60 percent while retaining low cost materials and manufacturing techniques. They can exceed the theoretical solar conversion efficiency limit for a single energy threshold material, that was calculated in 1961 by Shockley and Queisser as 31 percent under 1 sun illumination and 40.8 percent under maximal concentration of sunlight (46,200 suns, which makes the latter limit more difficult to approach than the former).

There are a few approaches to achieving these high efficiencies:

- Multijunction photovoltaic cell (multiple energy threshold devices).
- Modifying incident spectrum (concentration).
- Use of excess thermal generation (caused by UV light) to enhance voltages or carrier collection.
- Use of infrared spectrum to produce electricity at night.

Technologies includes:

- Silicon nanostructures.
- Up/Down converters.
- Hot-carrier cells.
- Thermoelectric cells.

High Efficiency Cells

High efficiency solar cells are a class of solar cells that can generate electricity at higher efficiencies than conventional solar cells. While high efficiency solar cells are more efficient in terms of electrical output per incident energy (watt/watt), much of the industry is focused on the most cost efficient technologies (cost-per-watt or $/watt). Still, many businesses and academics are focused on increasing the electrical efficiency of cells, and much development is focused on high efficiency solar cells.

Applications and Implementations

Polycrystaline PV cells laminated to backing material in a PV module.

Solar cells are often electrically connected and encapsulated as a module. PV modules often have a sheet of glass on the front (sun up) side, allowing light to pass while protecting the semiconductor wafers from the elements (rain, hail, etc.). Solar cells are also usually connected in series in modules, creating an additive voltage. Connecting cells in parallel will yield a higher current. Modules are then interconnected, in series or parallel, or both, to create an array with the desired peak DC voltage and current.

The power output of a solar array is measured in watts or kilowatts. In order to calculate the typical energy needs of the application, a measurement in watt-hours, kilowatt-hours or kilowatt-hours per day is often used. A common rule of thumb is that average power is equal to 20 percent of peak power, so that each peak kilowatt of solar array output power corresponds to energy production of 4.8 kWh per day (24 hours x 1kWh x 20 percent = 4.8 kWh)

To make practical use of the solar-generated energy, the electricity is most often fed into the electricity grid using inverters (grid-connected PV systems); in stand alone systems, batteries are used to store the energy that is not needed immediately.

Theory

Simple Explanation

- Photons in sunlight hit the solar panel and are absorbed by semiconducting materials, such as silicon.

- Electrons (negatively charged) are knocked loose from their atoms, allowing them to flow through the material to produce electricity. Due to the special composition of solar cells, the electrons are only allowed to move in a single direction. The complementary positive charges that are also created (like bubbles) are called holes and flow in the direction opposite of the electrons in a silicon solar panel.
- An array of solar cells converts solar energy into a usable amount of direct current (DC) electricity.

Photogeneration of Charge Carriers

When a photon hits a piece of silicon, one of three things can happen:

- The photon can pass straight through the silicon—this (generally) happens for lower energy photons.
- The photon can reflect off the surface.
- The photon can be absorbed by the silicon, if the photon energy is higher than the silicon band gap value. This generates an electron-hole pair and sometimes heat, depending on the band structure.

When a photon is absorbed, its energy is given to an electron in the crystal lattice. Usually this electron is in the valence band, and is tightly bound in covalent bonds between neighboring atoms, and hence unable to move far. The energy given to it by the photon "excites" it into the conduction band, where it is free to move around within the semiconductor. The covalent bond that the electron was previously a part of now has one fewer electron—this is known as a hole. The presence of a missing covalent bond allows the bonded electrons of neighboring atoms to move into the "hole," leaving another hole behind, and in this way a hole can move through the lattice. Thus, it can be said that photons absorbed in the semiconductor create mobile electron-hole pairs.

A photon need only have greater energy than that of the band gap in order to excite an electron from the valence band into the conduction band. However, the solar frequency spectrum approximates a black body spectrum at ~6000 K, and as such, much of the solar radiation reaching the Earth is composed of photons with energies greater than the band gap of silicon. These higher energy photons will be absorbed by the solar cell, but the difference in energy between these photons and the silicon band gap is converted into heat (via lattice vibrations—called phonons) rather than into usable electrical energy.

Charge Carrier Separation

There are two main modes for charge carrier separation in a solar cell:

- Drift of carriers, driven by an electrostatic field established across the device.
- Diffusion of carriers from zones of high carrier concentration to zones of low carrier concentration (following a gradient of electrochemical potential).

In the widely used p-n junction solar cells, the dominant mode of charge carrier separation is by drift. However, in non-p-n-junction solar cells (typical of the third generation solar cell research such as dye and polymer solar cells), a general electrostatic field has been confirmed to be absent, and the dominant mode of separation is via charge carrier diffusion.

The p-n Junction

The most commonly known solar cell is configured as a large-area p-n junction made from silicon. As a simplification, one can imagine bringing a layer of n-type silicon into direct contact with a layer of p-type silicon. In practice, p-n junctions of silicon solar cells are not made in this way, but rather, by diffusing an n-type dopant into one side of a p-type wafer (or vice versa).

If a piece of p-type silicon is placed in intimate contact with a piece of n-type silicon, then a diffusion of electrons occurs from the region of high electron concentration (the n-type side of the junction) into the region of low electron concentration (p-type side of the junction). When the electrons diffuse across the p-n junction, they recombine with holes on the p-type side. The diffusion of carriers does not happen indefinitely however, because of an electric field which is created by the imbalance of charge immediately on either side of the junction which this diffusion creates. The electric field established across the p-n junction creates a diode that promotes current in only one direction across the junction. Electrons may pass from the n-type side into the p-type side, and holes may pass from the p-type side to the n-type side, but not the other way around. This region where electrons have diffused across the junction is called the depletion region because it no longer contains any mobile charge carriers. It is also known as the "space charge region."

Connection to an External Load

Ohmic metal-semiconductor contacts are made to both the n-type and p-type sides of the solar cell, and the electrodes connected to an external load. Electrons that are created on the n-type side, or have been "collected" by the junction and swept onto the n-type side, may travel through the wire, power the load, and continue through the wire until they reach the p-type semiconductor-metal contact. Here, they recombine with a hole that was either created as an electron-hole pair on the p-type side of the solar cell, or are swept across the junction from the n-type side after being created there.

The voltage measured is equal to the difference in the quasi Fermi levels of the minority carriers, i.e., electrons in the p-type portion, and holes in the n-type portion.

Equivalent Circuit of a Solar Cell

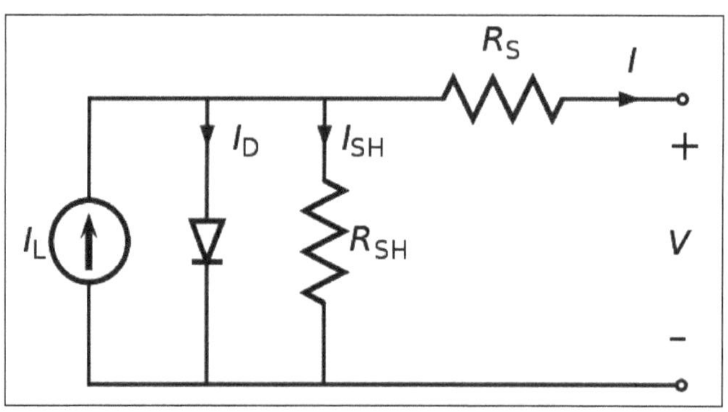

The equivalent circuit of a solar cell.

To understand the electronic behavior of a solar cell, it is useful to create a model which is electrically equivalent, and is based on discrete electrical components whose behavior is well known. An ideal solar cell may be modeled by a current source in parallel with a diode; in practice no solar cell is ideal, so a shunt resistance and a series resistance component are added to the model. The resulting equivalent circuit of a solar cell is shown on the left. Also shown, on the right, is the schematic representation of a solar cell for use in circuit diagrams.

Characteristic Equation

From the equivalent circuit it is evident that the current produced by the solar cell is equal to that produced by the current source, minus that which flows through the diode, minus that which flows through the shunt resistor:

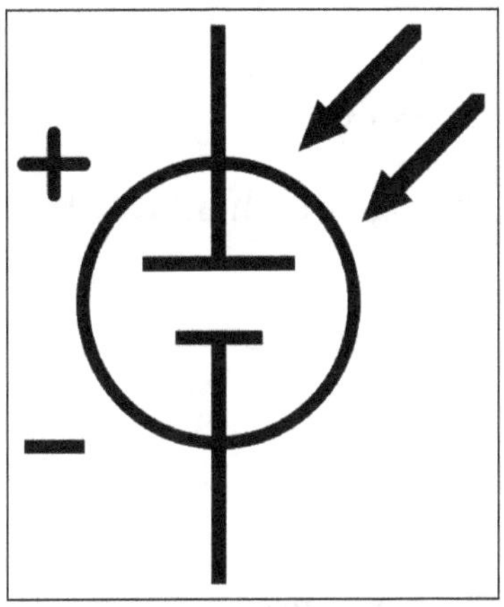

The schematic symbol of a solar cell.

$$I = I_L - I_D - I_{SH}$$

where

- I = output current (amperes)
- I_L = photogenerated current (amperes)
- I_D = diode current (amperes)
- I_{SH} = shunt current (amperes)

The current flowing through these elements governed by the voltage across them:

$$V_j = V + IR_S$$

where

- V = voltage across the output terminals (volts)
- I = output current (amperes)
- R_S = series resistance (Ω)

By the Shockley diode equation, the current diverted through the diode is:

$$I_D = I_0 \left\{ \exp\left[\frac{qV_j}{nkT} \right] - 1 \right\}$$

where

- I_0 = reverse saturation current (amperes)
- n = diode ideality factor (1 for an ideal diode)
- q = elementary charge
- k = Boltzmann's constant
- T = absolute temperature
- For silicon at 25°C, $kT / q \approx 0.0259$ volts.

By Ohm's law, the current diverted through the shunt resistor is:

$$I_{SH} = \frac{V_j}{R_{SH}}$$

where

- R_{SH} = shunt resistance (Ω)

Substituting these into the first equation produces the characteristic equation of a solar cell, which relates solar cell parameters to the output current and voltage:

$$I = I_L - I_0 \left\{ \exp\left[\frac{q(V + IR_S)}{nkT} \right] - 1 \right\} - \frac{V + IR_S}{R_{SH}}$$

An alternative derivation produces an equation similar in appearance, but with V on the left-hand side. The two alternatives are identities; that is, they yield precisely the same results.

In principle, given a particular operating voltage V the equation may be solved to determine the operating current I at that voltage. However, because the equation involves I on both sides in a transcendental function the equation has no general analytical solution. However, even without a solution it is physically instructive. Furthermore, it is easily solved using numerical methods. (A general analytical solution to the equation is possible using Lambert's W function, but since Lambert's W generally itself must be solved numerically this is a technicality.)

Since the parameters I_0, n, R_S, and R_{SH} cannot be measured directly, the most common application of the characteristic equation is nonlinear regression to extract the values of these parameters on the basis of their combined effect on solar cell behavior.

Effect of Physical Size

The values of I_0, R_S, and R_{SH} are dependent upon the physical size of the solar cell. In comparing otherwise identical cells, a cell with twice the surface area of another will, in principle, have double the I_0 because it has twice the junction area across which current can leak. It will also have half the R_S and R_{SH} because it has twice the cross-sectional area through which current can flow. For this reason, the characteristic equation is frequently written in terms of current density, or current produced per unit cell area:

$$J = J_L - J_0\left\{\exp\left[\frac{q(V + Jr_S)}{nkT}\right] - 1\right\} - \frac{V + Jr_S}{r_{SH}}$$

where

- J = current density (amperes/cm^2)
- J_L = reverse saturation current density (amperes/cm^2)
- r_S = specific series resistance (Ω-cm^2)
- r_{SH} = specific shunt resistance (Ω-cm^2)

This formulation has several advantages. One is that since cell characteristics are referenced to a common cross-sectional area they may be compared for cells of different physical dimensions. While this is of limited benefit in a manufacturing setting, where all cells tend to be the same size, it is useful in research and in comparing cells between manufacturers. Another advantage is that the density equation naturally scales the parameter values to similar orders of magnitude, which can make numerical extraction of them simpler and more accurate even with naive solution methods.

A practical limitation of this formulation is that as cell sizes shrink, certain parasitic effects grow in importance and can affect the extracted parameter values. For example, recombination and contamination of the junction tend to be greatest at the perimeter of the cell, so very small cells may exhibit higher values of J_0 or lower values of r_{SH} than larger cells that are otherwise identical. In such cases, comparisons between cells must be made cautiously and with these effects in mind.

Cell Temperature

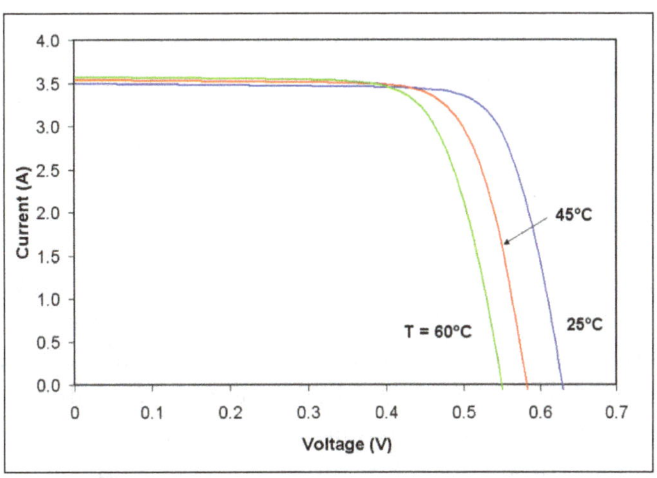

Effect of temperature on the current-voltage characteristics of a solar cell.

Temperature affects the characteristic equation in two ways: directly, via T in the exponential term, and indirectly via its effect on I_0. (Strictly speaking, temperature affects all of the terms, but these two far more significantly than the others.) While increasing reduces the magnitude of the exponent in the characteristic equation, the value of I_0 increases in proportion to $\exp T$. The net effect is to reduce Voc linearly with increasing temperature. The magnitude of this reduction is inversely proportional to Voc; that is, cells with higher values of Voc suffer smaller reductions in voltage with increasing temperature. For most crystalline silicon solar cells the reduction is about 0.50%/°C, though the rate for the highest-efficiency crystalline silicon cells is around 0.35%/°C. By way of comparison, the rate for amorphous silicon solar cells is 0.20-0.30%/°C, depending on how the cell is made.

The amount of photogenerated current I_L increases slightly with increasing temperature because of an increase in the number of thermally generated carriers in the cell. This effect is slight, however: about 0.065%/°C for crystalline silicon cells and 0.09 percent for amorphous silicon cells.

The overall effect of temperature on cell efficiency can be computed using these factors in combination with the characteristic equation. However, since the change in voltage is much stronger than the change in current, the overall effect on efficiency tends to be similar to that on voltage. Most crystalline silicon solar cells decline in efficiency by 0.50%/°C and most amorphous cells decline by 0.15-0.25%/°C. The figure to the right shows I-V curves that might typically be seen for a crystalline silicon solar cell at various temperatures.

Series Resistance

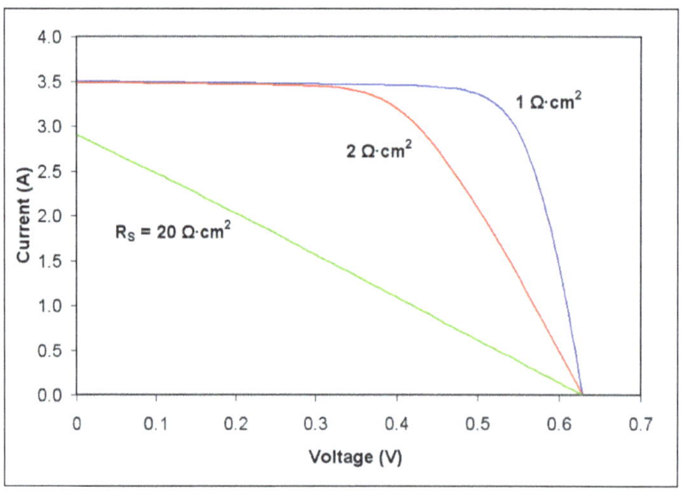

Effect of series resistance on the current-voltage characteristics of a solar cell.

As series resistance increases, the voltage drop between the junction voltage and the terminal voltage becomes greater for the same flow of current. The result is that the current-controlled portion of the I-V curve begins to sag toward the origin, producing a significant decrease in the terminal voltage V and a slight reduction in I_{sc}. Very high values of R_s will also produce a significant reduction in I_{sc}; in these regimes, series resistance dominates and the behavior of the solar cell resembles that of a resistor. These effects are shown for crystalline silicon solar cells in the I-V curves displayed in the figure to the right.

Shunt Resistance

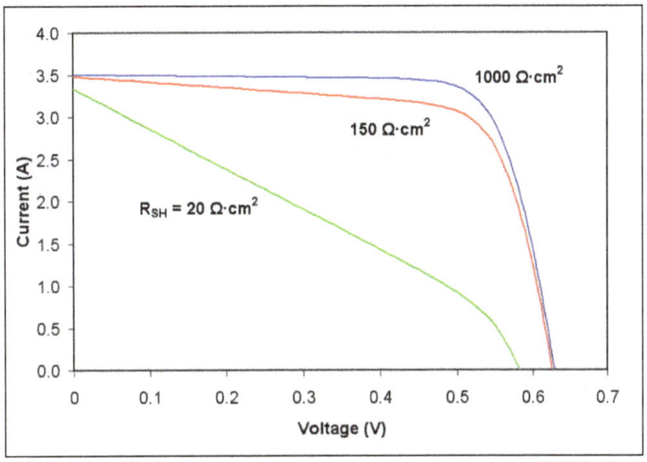

Effect of shunt resistance on the current-voltage characteristics of a solar cell.

As shunt resistance decreases, the flow of current diverted through the shunt resistor increases for a given level of junction voltage. The result is that the voltage-controlled portion of the I-V curve begins to sag toward the origin, producing a significant decrease in the terminal current I and a slight reduction in Voc. Very low values of R_{SH} will produce a significant reduction in Voc. Much as in the case of a high series resistance, a badly shunted solar cell will take on operating characteristics similar to those of a resistor. These effects are shown for crystalline silicon solar cells in the I-V curves displayed in the figure to the right.

Reverse Saturation Current

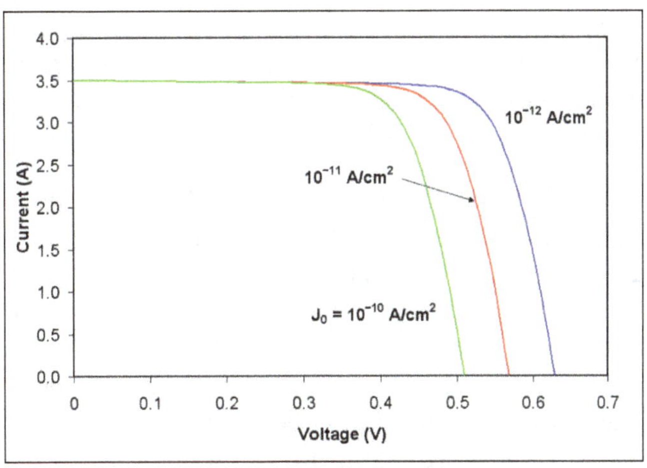

Effect of reverse saturation current on the current-voltage characteristics of a solar cell.

If one assumes infinite shunt resistance, the characteristic equation can be solved for Voc:

$$V_{OC} = \frac{kT}{q}\exp\left\{\frac{I_{SC}}{I_0}+1\right\}.$$

Thus, an increase in I_0 produces a reduction in Voc proportional to the inverse of the logarithm of the increase. This explains mathematically the reason for the reduction in Voc that accompanies increases in temperature described above. The effect of reverse saturation current on the I-V curve of a crystalline silicon solar cell are shown in the figure to the right. Physically, reverse saturation current is a measure of the "leakage" of carriers across the p-n junction in reverse bias. This leakage is a result of carrier recombination in the neutral regions on either side of the junction.

Ideality Factor

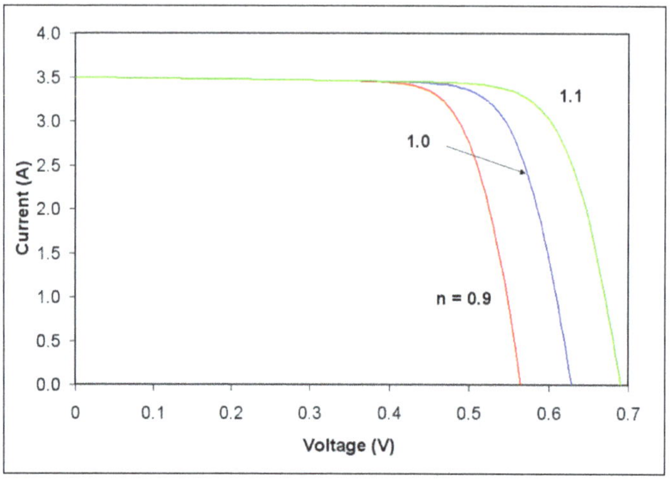

Effect of ideality factor on the current-voltage characteristics of a solar cell.

The ideality factor (also called the emissivity factor) is a fitting parameter that describes how closely the diode's behavior matches that predicted by theory, which assumes the p-n junction of the diode is an infinite plane and no recombination occurs within the space-charge region. A perfect match to theory is indicated when $n = 1$. When recombination in the space-charge region dominate other recombination, however, $n = 2$. The effect of changing ideality factor independently of all other parameters is shown for a crystalline silicon solar cell in the I-V curves displayed in the figure to the right.

Most solar cells, which are quite large compared to conventional diodes, well approximate an infinite plane and will usually exhibit near-ideal behavior under Standard Test Condition ($n \approx 1$). Under certain operating conditions, however, device operation may be dominated by recombination in the space-charge region. This is characterized by a significant increase in I_0 as well as an increase in ideality factor to $n \approx 2$. The latter tends to erode solar cell output voltage while the former acts to increase it. The net effect, therefore, is a combination of the increase in voltage shown for increasing n in the figure to the right and the decrease in voltage shown for increasing I_0 in the figure above. Typically, I_0 is the more significant factor and the result is a reduction in voltage.

Solar Cell Efficiency Factors

Energy Conversion Efficiency

A solar cell's energy conversion efficiency (η, "eta"), is the percentage of power converted (from absorbed light to electrical energy) and collected, when a solar cell is connected to an electrical circuit. This term is calculated using the ratio of the maximum power point, P_m, divided by the input light irradiance (E, in W/m²) under standard test conditions (STC) and the surface area of the solar cell (A_c in m²).

$$\eta = \frac{P_m}{E \times A_c}$$

STC specifies a temperature of 25°C and an irradiance of 1000 W/m² with an air mass 1.5 (AM1.5) spectrum. These correspond to the irradiance and spectrum of sunlight incident on a clear day upon a sun-facing 37°-tilted surface with the sun at an angle of 41.81° above the horizon. This condition approximately represents solar noon near the spring and autumn equinoxes in the continental United States with surface of the cell aimed directly at the sun. Thus, under these conditions a solar cell of 12 percent efficiency with a 100 cm² (0.01 m²) surface area can be expected to produce approximately 1.2 watts of power.

The losses of a solar cell may be broken down into reflectance losses, thermodynamic efficiency, recombination losses and resistive electrical loss. The overall efficiency is the product of each of these individual losses.

Due to the difficulty in measuring these parameters directly, other parameters are measured instead: Thermodynamic Efficiency, Quantum Efficiency, V_{OC} ratio, and Fill Factor. Reflectance losses are a portion of the Quantum Efficiency under "External Quantum Efficiency." Recombination losses make up a portion of the Quantum Efficiency, V_{OC} ratio, and Fill Factor. Resistive losses are predominantly categorized under Fill Factor, but also make up minor portions of the Quantum Efficiency, V_{OC} ratio.

Generally, solar cells on the market today do not produce much electricity from ultraviolet light, instead it is either filtered out or absorbed by the cell, heating the cell. That heat is wasted energy and could even lead to damage to the cell.

Thermodynamic Efficiency Limit

Solar cells operate as quantum energy conversion devices, and are therefore subject to the "Thermodynamic Efficiency Limit." Photons with an energy below the band gap of the absorber material cannot generate a hole-electron pair, and so their energy is not converted to useful output and only generates heat if absorbed. For photons with an energy above the band gap energy, only a fraction of the energy above the band gap can be converted to useful output. When a photon of greater energy is absorbed, the excess energy above the band gap is converted to kinetic energy of the carrier combination. The excess kinetic energy is converted to heat through phonon interactions as the kinetic energy of the carriers slows to equilibrium velocity.

Solar cells with multiple band gap absorber materials are able to more efficiently convert the solar spectrum. By using multiple band gaps, the solar spectrum may be broken down into smaller bins where the thermodynamic efficiency limit is higher for each bin.

Quantum Efficiency

When a photon is absorbed by a solar cell it can produce a pair of free charge carriers, i.e., an electron-hole pair. One of the carriers (the minority carrier) may then be able to reach the p-n junction and contribute to the current produced by the solar cell; such a carrier is said to be collected. Alternatively, the carrier may give up its energy and once again become bound to an atom within the solar cell without being collected; this process is then called recombination since one electron and one hole recombine and thereby annihilate the associated free charge. The carriers that recombine do not contribute to the generation of electrical current.

Quantum efficiency refers to the percentage of photons that are converted to electric current (i.e., collected carriers) when the cell is operated under short circuit conditions. External quantum efficiency (EQE) is the fraction of incident photons that are converted to electrical current, while internal quantum efficiency (IQE) is the fraction of absorbed photons that are converted to electrical current. Mathematically, internal quantum efficiency is related to external quantum efficiency by the reflectance (R) and the transmittance (T) of the solar cell by $IQE = EQE / (1 - R - T)$. Please note that for a thick bulk Si solar cell T is approximately zero and is therefore in practical cases often neglected.

Quantum efficiency should not be confused with energy conversion efficiency, as it does not convey information about the fraction of power that is converted by the solar cell. Furthermore, quantum efficiency is most usefully expressed as a spectral measurement (that is, as a function of photon wavelength or energy). Since some wavelengths are absorbed more effectively than others in most semiconductors, spectral measurements of quantum efficiency can yield valuable information about which parts of a particular solar cell design are most in need of improvement. Spectral quantum efficiency measurements also allow determining important solar cell parameters such as the base diffusion length.

Maximum-power Point

A solar cell may operate over a wide range of voltages (V) and currents (I). By increasing the resistive load on an irradiated cell continuously from zero (a short circuit) to a very high value (an open circuit) one can determine the maximum-power point, the point that maximizes V×I; that is, the load for which the cell can deliver maximum electrical power at that level of irradiation. (The output power is zero in both the short circuit and open circuit extremes).

A high quality, monocrystalline silicon solar cell, at 25 °C cell temperature, may produce 0.60 volts open-circuit (Voc). The cell temperature in full sunlight, even with 25 °C air temperature, will probably be close to 45 °C, reducing the open-circuit voltage to 0.55 volts per cell. The voltage drops modestly, with this type of cell, until the short-circuit current is approached (Isc). Maximum power (with 45 °C cell temperature) is typically produced with 75 percent to 80 percent of the open-circuit voltage (0.43 volts in this case) and 90 percent of the short-circuit current. This output can be up to 70 percent of the Voc x Isc product. The short-circuit current (Isc) from a cell is nearly proportional to the illumination, while the open-circuit voltage (Voc) may drop only 10 percent with a 80 percent drop in illumination. Lower-quality cells have a more rapid drop in voltage with increasing current and could produce only 1/2 Voc at 1/2 Isc. The usable power output could thus drop from 70 percent of the Voc x Isc product to 50 percent or even as little as 25 percent. Vendors who rate their solar cell "power" only as Voc x Isc, without giving load curves, can be seriously distorting their actual performance.

The maximum power point of a photovoltaic varies with incident illumination. For systems large enough to justify the extra expense, a maximum power point tracker tracks the instantaneous power by continually measuring the voltage and current (and hence, power transfer), and uses this information to dynamically adjust the load so the maximum power is always transferred, regardless of the variation in lighting.

Fill factor

Another defining term in the overall behavior of a solar cell is the fill factor (FF). This is the ratio of the maximum power point divided by the open circuit voltage (V_{oc}) and the short circuit current (I_{sc}):

$$FF = \frac{P_m}{V_{oc} \times I_{sc}} = \frac{\eta \times A_c \times E}{V_{oc} \times I_{sc}}.$$

Comparison of Energy Conversion Efficiencies

At this point, discussion of the different ways to calculate efficiency for space cells and terrestrial cells is necessary to alleviate confusion. In space, where there is no atmosphere, the spectrum of the sun is relatively unfiltered. However, on earth, with air filtering the incoming light, the solar spectrum changes. To account for the spectral differences, a system was devised to calculate this filtering effect. Simply, the filtering effect ranges from Air Mass 0 (AM0) in space, to approximately Air Mass 1.5 on earth. Multiplying the spectral differences by the quantum efficiency of the solar cell in question will yield the efficiency of the device. For example, a Silicon solar cell in space might have an efficiency of 14 percent at AM0, but have an efficiency of 16 percent on earth at AM 1.5. Terrestrial efficiencies typically are greater than space efficiencies.

Solar cell efficiencies vary from 6 percent for amorphous silicon-based solar cells to 40.7 percent with multiple-junction research lab cells and 42.8 percent with multiple dies assembled into a hybrid package. Solar cell energy conversion efficiencies for commercially available multicrystalline Si solar cells are around 14-19 percent. The highest efficiency cells have not always been the most economical—for example a 30 percent efficient multijunction cell based on exotic materials such as gallium arsenide or indium selenide and produced in low volume might well cost one hundred times as much as an 8% efficient amorphous silicon cell in mass production, while only delivering about four times the electrical power.

However, there is a way to "boost" solar power. By increasing the light intensity, typically photogenerated carriers are increased, resulting in increased efficiency by up to 15 percent. These so-called "concentrator systems" have only begun to become cost-competitive as a result of the development of high efficiency GaAs cells. The increase in intensity is typically accomplished by using concentrating optics. A typical concentrator system may use a light intensity 6-400 times the sun, and increase the efficiency of a one sun GaAs cell from 31 percent at AM 1.5 to 35 percent.

A common method used to express economic costs of electricity-generating systems is to calculate a price per delivered kilowatt-hour (kWh). The solar cell efficiency in combination with the available irradiation has a major influence on the costs, but generally speaking the overall system efficiency is important. Using the commercially available solar cells (as of 2006) and system technology leads to system efficiencies between 5 and 19 percent. As of 2005, photovoltaic electricity generation costs ranged from ~0.60 US$/kWh (0.50 €/kWh) (central Europe) down to ~0.30 US$/kWh (0.25 €/kwh in regions of high solar irradiation. This electricity is generally fed into the electrical grid on the customer's side of the meter. The cost can be compared to prevailing retail electric pricing (as of 2005), which varied from between 0.04 and 0.50 US$/kWh worldwide. (In addition to solar irradiance profiles, these costs/kwh calculations will vary depending on assumptions for years of useful life of a system. Most c-Si panels are warranted for 25 years and should see 35+ years of useful life.)

The chart below illustrates the various commercial large-area module energy conversion efficiencies and the best laboratory efficiencies obtained for various materials and technologies.

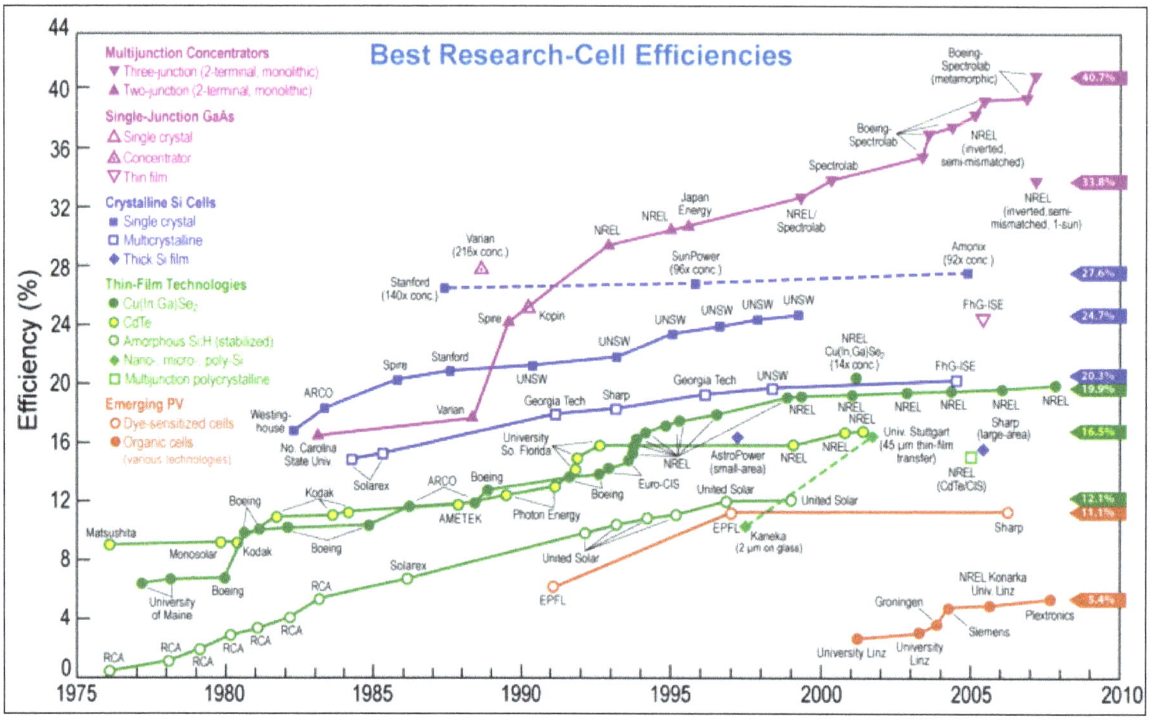

Reported timeline of solar cell energy conversion efficiencies
(from National Renewable Energy Laboratory (USA).

Watts Peak

Since solar cell output power depends on multiple factors, such as the sun's incidence angle, for comparison purposes between different cells and panels, the measure of watts peak (Wp) is used. It is the output power under these conditions known as STC. Larger panels use the same rating system, but use kWp (1000 watts peak).

- Insolation (solar irradiance) 1000 W/m²
- Solar reference spectrum AM (airmass) 1.5
- Cell temperature 25 °C

Solar Cells and Energy Payback

Energy payback is the recovery (period) of the energy spent for manufacturing of the respective technical energy systems, also called harvesting ratio (ISO 13602).

In the 1990s, when silicon cells were twice as thick, efficiencies were 30 percent lower than today and lifetimes were shorter, it may well have cost more energy to make a cell than it could generate in a lifetime. In the meantime, the technology has progressed significantly, and the energy payback time of a modern photovoltaic module is typically from 1 to 4 years depending on the type and where it is used. Generally, thin film technologies - despite having comparatively low conversion efficiencies - achieve significantly shorter energy payback times than conventional systems (often < 1 year). With a typical lifetime of 20 to 30 years, this means that modern solar cells are net energy producers, i.e., they generate significantly more energy over their lifetime than the energy expended in producing them.

Light-absorbing Materials

All solar cells require a light absorbing material contained within the cell structure to absorb photons and generate electrons via the photoelectric effect. The materials used in solar cells tend to have the property of preferentially absorbing the wavelengths of solar light that reach the earth surface; however, some solar cells are optimized for light absorption beyond Earth's atmosphere as well. Light absorbing materials can often be used in multiple physical configurations to take advantage of different light absorption and charge separation mechanisms.

Photovoltaic panels are normally made of either silicon or thin-film cells.

Many currently available solar cells are configured as bulk materials that are subsequently cut into wafers and treated in a "top-down" method of synthesis (silicon being the most prevalent bulk material).

Other materials are configured as thin-films (inorganic layers, organic dyes, and organic polymers) that are deposited on supporting substrates, while a third group are configured as nanocrystals and used as quantum dots (electron-confined nanoparticles) embedded in a supporting matrix in a "bottom-up" approach. Silicon remains the only material that is well-researched in both bulk (also called wafer-based) and thin-film configurations.

There are many new alternatives to Silicon photocells. Proprietary nanoparticle silicon printing processes promises many of the photovoltaic features that conventional silicon can never achieve. It can be printed reel-to-reel on stainless steel or other high temperature substrates. However, most of the work on the next generation of photovoltaics is directed at printing onto low cost flexible polymer film and ultimately on common packaging materials. The main contenders are currently CIGS, CdTe, DSSC and organic photovoltaics.

Thin-film Solar Cell

A thin-film solar cell is a second generation solar cell that is made by depositing one or more thin layers, or thin film (TF) of photovoltaic material on a substrate, such as glass, plastic or metal. Thin-film solar cells are commercially used in several technologies, including cadmium telluride (CdTe), copper indium gallium diselenide (CIGS), and amorphous thin-film silicon (a-Si, TF-Si).

Film thickness varies from a few nanometers (nm) to tens of micrometers (μm), much thinner than thin-film's rival technology, the conventional, first-generation crystalline silicon solar cell (c-Si), that uses wafers of up to 200 μm thick. This allows thin film cells to be flexible, and lower in weight. It is used in building integrated photovoltaics and as semi-transparent, photovoltaic glazing material that can be laminated onto windows. Other commercial applications use rigid thin film solar panels (interleaved between two panes of glass) in some of the world's largest photovoltaic power stations.

Thin-film technology has always been cheaper but less efficient than conventional c-Si technology. However, it has significantly improved over the years. The lab cell efficiency for CdTe and CIGS is now beyond 21 percent, outperforming multicrystalline silicon, the dominant material currently used in most solar PV systems. Accelerated life testing of thin film modules under laboratory conditions measured a somewhat faster degradation compared to conventional PV, while a lifetime of 20 years or more is generally expected. Despite these enhancements, market-share of thin-film never reached more than 20 percent in the last two decades and has been declining in recent years to about 9 percent of worldwide photovoltaic installations in 2013.

Thin-film solar cells, a second generation of photovoltaic (PV) solar cells:

- Top: Thin-film silicon laminates being installed onto a roof.
- Middle: CIGS solar cell on a flexible plastic backing and rigid CdTe panels mounted on a supporting structure
- Bottom: thin-film laminates on rooftops

Other thin-film technologies that are still in an early stage of ongoing research or with limited commercial availability are often classified as emerging or third generation photovoltaic cells and include organic, and dye-sensitized, as well as quantum dot, copper zinc tin sulfide, nanocrystal, micromorph, and perovskite solar cells.

Materials

Thin-film technologies reduce the amount of active material in a cell. Most sandwich active material between two panes of glass. Since silicon solar panels only use one pane of glass, thin film panels are approximately twice as heavy as crystalline silicon panels, although they have a smaller ecological impact (determined from life cycle analysis). The majority of film panels have 2-3 percentage points lower conversion efficiencies than crystalline silicon. Cadmium telluride (CdTe), copper indium gallium selenide (CIGS) and amorphous silicon (a-Si) are three thin-film technologies often used for outdoor applications.

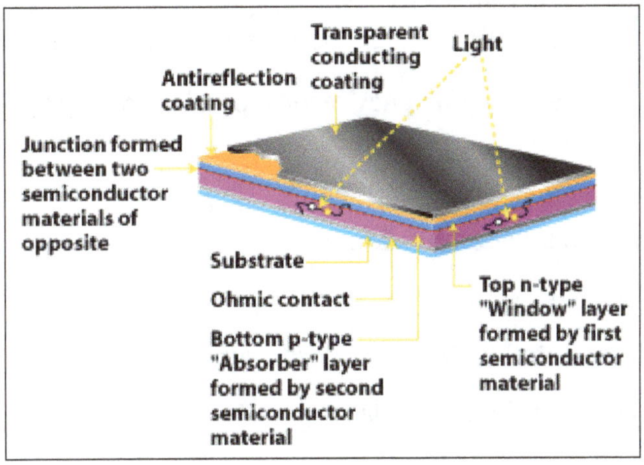

Cross-section of a TF cell.

Cadmium Telluride

Cadmium telluride (CdTe) is the predominant thin film technology. With about 5 percent of worldwide PV production, it accounts for more than half of the thin film market. The cell's lab efficiency has also increased significantly in recent years and is on a par with CIGS thin film and close to the efficiency of multi-crystalline silicon as of 2013. Also, CdTe has the lowest Energy payback time of all mass-produced PV technologies, and can be as short as eight months in favorable locations. A prominent manufacturer is the US-company First Solar based in Tempe, Arizona, that produces CdTe-panels with an efficiency of about 14 percent at a reported cost of $0.59 per watt.

Although the toxicity of cadmium may not be that much of an issue and environmental concerns completely resolved with the recycling of CdTe modules at the end of their life time, there are still uncertainties and the public opinion is skeptical towards this technology. The usage of rare materials may also become a limiting factor to the industrial scalability of CdTe thin film technology. The rarity of tellurium—of which telluride is the anionic form—is comparable to that of platinum in the earth's crust and contributes significantly to the module's cost.

Copper Indium Gallium Selenide

A copper indium gallium selenide solar cell or CIGS cell uses an absorber made of copper, indium, gallium, selenide (CIGS), while gallium-free variants of the semiconductor material are abbreviated CIS. It is one of three mainstream thin-film technologies, the other two being cadmium telluride and amorphous silicon, with a lab-efficiency above 20 percent and a share of 2 percent in the overall PV market in 2013. A prominent manufacturer of cylindrical CIGS-panels was the now-bankrupt company Solyndra in Fremont, California. Traditional methods of fabrication involve vacuum processes including co-evaporation and sputtering. In 2008, IBM and Tokyo Ohka Kogyo Co., Ltd. (TOK) announced they had developed a new, non-vacuum, solution-based manufacturing process for CIGS cells and are aiming for efficiencies of 15% and beyond.

Hyperspectral imaging has been used to characterize these cells. Researchers from IRDEP (Institute of Research and Development in Photovoltaic Energy) in collaboration with Photon etc. were able to determine the splitting of the quasi-Fermi level with photoluminescence mapping while the electroluminescence data were used to derive the external quantum efficiency (EQE). Also, through a light beam induced current (LBIC) cartography experiment, the EQE of a microcrystalline CIGS solar cell could be determined at any point in the field of view.

As of April 2019, current conversion efficiency record for a laboratory CIGS cell stands at 22.9%.

Silicon

Three major silicon-based module designs dominate:

- Amorphous silicon cells.
- Amorphous / microcrystalline tandem cells (micromorph).
- Thin-film polycrystalline silicon on glass.

Amorphous Silicon

Amorphous silicon (a-Si) is a non-crystalline, allotropic form of silicon and the most well-developed thin film technology to-date. Thin-film silicon is an alternative to conventional *wafer* (or *bulk*) crystalline silicon. While chalcogenide-based CdTe and CIS thin films cells have been developed in the lab with great success, there is still industry interest in silicon-based thin film cells. Silicon-based devices exhibit fewer problems than their CdTe and CIS counterparts such as toxicity and humidity issues with CdTe cells and low manufacturing yields of CIS due to material complexity. Additionally, due to political resistance to the use non-"green" materials in solar energy production, there is no stigma in the use of standard silicon.

This type of thin-film cell is mostly fabricated by a technique called plasma-enhanced chemical vapor deposition. It uses a gaseous mixture of silane (SiH_4) and hydrogen to deposit a very thin layer of only 1 micrometre (μm) of silicon on a substrate, such as glass, plastic or metal, that has already been coated with a layer of transparent conducting oxide. Other methods used to deposit amorphous silicon on a substrate include sputtering and hot wire chemical vapor deposition techniques.

a-Si is attractive as a solar cell material because it's an abundant, non-toxic material. It requires a low processing temperature and enables a scalable production upon a flexible, low-cost substrate with little silicon material required. Due to its bandgap of 1.7 eV, amorphous silicon also absorbs a very broad range of the light spectrum, that includes infrared and even some ultraviolet and performs very well at weak light. This allows the cell to generate power in the early morning, or late afternoon and on cloudy and rainy days, contrary to crystalline silicon cells, that are significantly less efficient when exposed at diffuse and indirect daylight.

However, the efficiency of an a-Si cell suffers a significant drop of about 10 to 30 percent during the first six months of operation. This is called the Staebler-Wronski effect (SWE) – a typical loss in electrical output due to changes in photoconductivity and dark conductivity caused by prolonged exposure to sunlight. Although this degradation is perfectly reversible upon annealing at or above 150 °C, conventional c-Si solar cells do not exhibit this effect in the first place.

Its basic electronic structure is the p-i-n junction. The amorphous structure of a-Si implies high inherent disorder and dangling bonds, making it a bad conductor for charge carriers. These dangling bonds act as recombination centers that severely reduce carrier lifetime. A p-i-n structure is usually used, as opposed to an n-i-p structure. This is because the mobility of electrons in a-Si:H is roughly 1 or 2 orders of magnitude larger than that of holes, and thus the collection rate of electrons moving from the n- to p-type contact is better than holes moving from p- to n-type contact. Therefore, the p-type layer should be placed at the top where the light intensity is stronger, so that the majority of the charge carriers crossing the junction are electrons.

Tandem-cell using a-Si/μc-Si

A layer of amorphous silicon can be combined with layers of other allotropic forms of silicon to produce a multi-junction solar cell. When only two layers (two p-n junctions) are combined, it is called a tandem-cell. By stacking these layers on top of one other, a broader range of the light spectra is absorbed, improving the cell's overall efficiency.

In micromorphous silicon, a layer of microcrystalline silicon (μc-Si) is combined with amorphous silicon, creating a tandem cell. The top a-Si layer absorbs the visible light, leaving the infrared part to the bottom μc-Si layer. The micromorph stacked-cell concept was pioneered and patented at the Institute of Microtechnology (IMT) of the Neuchâtel University in Switzerland, and was licensed to TEL Solar. A new world record PV module based on the micromorph concept with 12.24% module efficiency was independently certified in July 2014.

Because all layers are made of silicon, they can be manufactured using PECVD. The band gap of a-Si is 1.7 eV and that of c-Si is 1.1 eV. The c-Si layer can absorb red and infrared light. The best efficiency can be achieved at transition between a-Si and c-Si. As nanocrystalline silicon (nc-Si) has about the same bandgap as c-Si, nc-Si can replace c-Si.

Tandem-cell using a-Si/pc-Si

Amorphous silicon can also be combined with protocrystalline silicon (pc-Si) into a tandem-cell. Protocrystalline silicon with a low volume fraction of nanocrystalline silicon is optimal for high open-circuit voltage. These types of silicon present dangling and twisted bonds, which results in deep defects (energy levels in the bandgap) as well as deformation of the valence and conduction bands (band tails).

Polycrystalline Silicon on Glass

A new attempt to fuse the advantages of bulk silicon with those of thin-film devices is thin film polycrystalline silicon on glass. These modules are produced by depositing an antireflection coating and doped silicon onto textured glass substrates using plasma-enhanced chemical vapor deposition (PECVD). The texture in the glass enhances the efficiency of the cell by approximately 3% by reducing the amount of incident light reflecting from the solar cell and trapping light inside the solar cell. The silicon film is crystallized by an annealing step, temperatures of 400–600 Celsius, resulting in polycrystalline silicon.

These new devices show energy conversion efficiencies of 8% and high manufacturing yields of >90%. Crystalline silicon on glass (CSG), where the polycrystalline silicon is 1–2 micrometres, is noted for its stability and durability; the use of thin film techniques also contributes to a cost savings over bulk photovoltaics. These modules do not require the presence of a transparent conducting oxide layer. This simplifies the production process twofold; not only can this step be skipped, but the absence of this layer makes the process of constructing a contact scheme much simpler. Both of these simplifications further reduce the cost of production. Despite the numerous advantages over alternative design, production cost estimations on a per unit area basis show that these devices are comparable in cost to single-junction amorphous thin film cells.

Gallium Arsenide

The semiconductor material gallium arsenide (GaAs) is also used for single-crystalline thin film solar cells. Although GaAs cells are very expensive, they hold the world record for the highest-efficiency, single-junction solar cell at 28.8%. GaAs is more commonly used in multi-junction solar cells for solar panels on spacecrafts, as the larger power to weight ratio lowers the launch costs in space-based solar power (InGaP/(In)GaAs/Ge cells). They are also used in concentrator photovoltaics, an emerging technology best suited for locations that receive much sunlight, using lenses to focus sunlight on a much smaller, thus less expensive GaAs concentrator solar cell.

Efficiencies

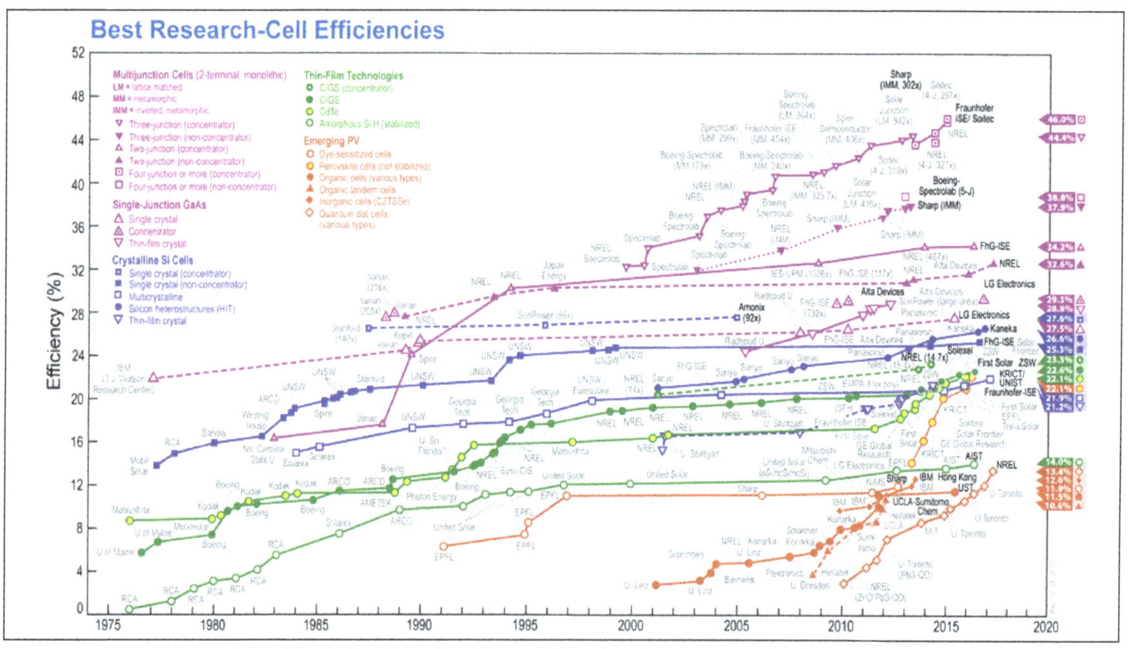

Solar cell efficiencies of various cell technologies as tracked by NREL.

Incremental improvements in efficiency began with the invention of the first modern silicon solar cell in 1954. By 2010 these steady improvements had resulted in modules capable of converting 12 to 18 percent of solar radiation into electricity. The improvements to efficiency have continued to accelerate in the years since 2010, as shown in the accompanying chart.

Cells made from newer materials tend to be less efficient than bulk silicon, but are less expensive to produce. Their quantum efficiency is also lower due to reduced number of collected charge carriers per incident photon.

The performance and potential of thin-film materials are high, reaching cell efficiencies of 12–20%; prototype module efficiencies of 7–13%; and production modules in the range of 9%. The thin film cell prototype with the best efficiency yields 20.4% (First Solar), comparable to the best conventional solar cell prototype efficiency of 25.6% from Panasonic.

NREL once predicted that costs would drop below $100/m^2 in volume production, and could later fall below $50/m^2.

A new record for thin film solar cell efficiency of 22.3% has been achieved by solar frontier the world's largest cis solar energy provider. In joint research with the New Energy and Industrial Technology Development Organization (NEDO) of Japan, Solar Frontier achieved 22.3% conversion efficiency on a 0.5 cm^2 cell using its CIS technology. This is an increase of 0.6 percentage points over the industry's previous thin-film record of 21.7%.

Absorption

Multiple techniques have been employed to increase the amount of light that enters the cell and reduce the amount that escapes without absorption. The most obvious technique is to minimizing the top contact coverage of the cell surface, reducing the area that blocks light from reaching the cell.

The weakly absorbed long wavelength light can be obliquely coupled into silicon and traverses the film several times to enhance absorption.

Multiple methods have been developed to increase absorption by reducing the number of incident photons being reflected away from the cell surface. An additional anti-reflective coating can cause destructive interference within the cell by modulating the refractive index of the surface coating. Destructive interference eliminates the reflective wave, causing all incident light to enter the cell.

Surface texturing is another option for increasing absorption, but increases costs. By applying a texture to the active material's surface, the reflected light can be refracted into striking the surface again, thus reducing reflectance.For example, black silicon texturing by reactive ion etching(RIE) is an effective and economic approach to increase the absorption of thin-film silicon solar cells. A textured backreflector can prevent light from escaping through the rear of the cell.

In addition to surface texturing, the plasmonic light-trapping scheme attracted a lot of attention to aid photocurrent enhancement in thin film solar cells. This method makes use of collective oscillation of excited free electrons in noble metal nanoparticles, which are influenced by particle shape, size and dielectric properties of the surrounding medium.

In addition to minimizing reflective loss, the solar cell material itself can be optimized to have higher chance of absorbing a photon that reaches it. Thermal processing techniques can significantly enhance the crystal quality of silicon cells and thereby increase efficiency. Layering thin-film cells to create a multi-junction solar cell can also be done. Each layer's band gap can be designed to best absorb a different range of wavelengths, such that together they can absorb a greater spectrum of light.

Further advancement into geometric considerations can exploit nanomaterial dimensionality. Large, parallel nanowire arrays enable long absorption lengths along the length of the wire while maintaining short minority carrier diffusion lengths along the radial direction. Adding nanoparticles between the nanowires allows conduction. The natural geometry of these arrays forms a textured surface that traps more light.

Multi-junction Solar Cell

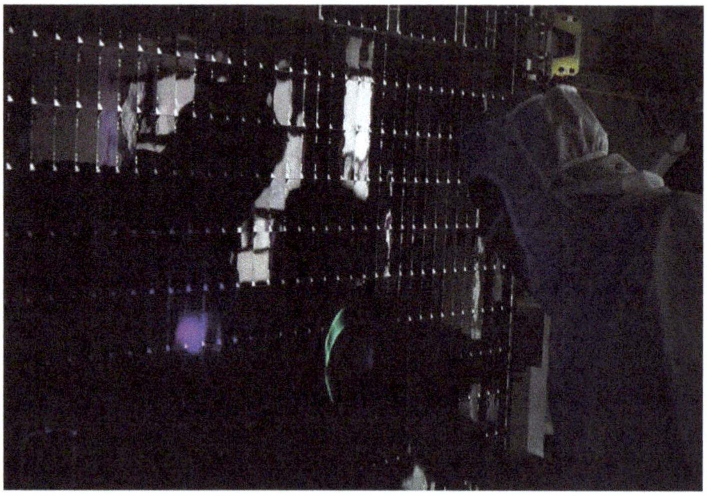

Black light test of Dawn's triple-junction gallium arsenide solar cells.

Multi-junction (MJ) solar cells are solar cells with multiple p–n junctions made of different semiconductor materials. Each material's p-n junction will produce electric current in response to different wavelengths of light. The use of multiple semiconducting materials allows the absorbance of a broader range of wavelengths, improving the cell's sunlight to electrical energy conversion efficiency.

Traditional single-junction cells have a maximum theoretical efficiency of 33.16%. Theoretically, an infinite number of junctions would have a limiting efficiency of 86.8% under highly concentrated sunlight.

Currently, the best lab examples of traditional crystalline silicon (c-Si) solar cells have efficiencies between 20% and 25%, while lab examples of multi-junction cells have demonstrated performance over 46% under concentrated sunlight. Commercial examples of tandem cells are widely available at 30% under one-sun illumination, and improve to around 40% under concentrated sunlight. However, this efficiency is gained at the cost of increased complexity and manufacturing price. To date, their higher price and higher price-to-performance ratio have limited their use to special roles, notably in aerospace where their high power-to-weight ratio is desirable. In terrestrial applications, these solar cells are emerging in concentrator photovoltaics (CPV), with a growing number of installations around the world.

Tandem fabrication techniques have been used to improve the performance of existing designs. In particular, the technique can be applied to lower cost thin-film solar cells using amorphous silicon, as opposed to conventional crystalline silicon, to produce a cell with about 10% efficiency that is lightweight and flexible. This approach has been used by several commercial vendors, but these products are currently limited to certain niche roles, like roofing materials.

Loss Mechanisms

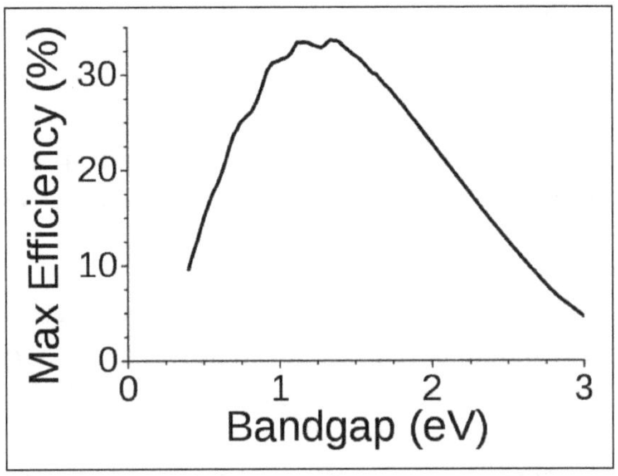

The Shockley-Queisser limit for the efficiency of a single-junction solar cell. It is essentially impossible for a single-junction solar cell, under unconcentrated sunlight, to have more than ~34% efficiency. A multi-junction cell, however, can exceed that limit.

The theoretical performance of a solar cell was first studied in depth in the 1960s, and is today known as the Shockley–Queisser limit. The limit describes several loss mechanisms that are inherent to any solar cell design.

The first are the losses due to blackbody radiation, a loss mechanism that affects any material object above absolute zero. In the case of solar cells at standard temperature and pressure, this loss accounts for about 7% of the power. The second is an effect known as "recombination", where the electrons created by the photoelectric effect meet the electron holes left behind by previous excitations. In silicon, this accounts for another 10% of the power.

However, the dominant loss mechanism is the inability of a solar cell to extract all of the power in the light, and the associated problem that it cannot extract any power at all from certain photons. This is due to the fact that the photons must have enough energy to overcome the bandgap of the material.

If the photon has less energy than the bandgap, it is not collected at all. This is a major consideration for conventional solar cells, which are not sensitive to most of the infrared spectrum, although that represents almost half of the power coming from the sun. Conversely, photons with more energy than the bandgap, say blue light, initially eject an electron to a state high above the bandgap, but this extra energy is lost through collisions in a process known as "relaxation". This lost energy turns into heat in the cell, which has the side-effect of further increasing blackbody losses.

Combining all of these factors, the maximum efficiency for a single-bandgap material, like conventional silicon cells, is about 34%. That is, 66% of the energy in the sunlight hitting the cell will be lost. Practical concerns further reduce this, notably reflection off the front surface or the metal terminals, with modern high-quality cells at about 22%.

Lower, also called narrower, bandgap materials will convert longer wavelength, lower energy photons. Higher, or wider bandgap materials will convert shorter wavelength, higher energy light. An analysis of the AM1.5 spectrum, shows the best balance is reached at about 1.1 eV (about 1100 nm, in the near infrared), which happens to be very close to the natural bandgap in silicon and a number of other useful semiconductors.

Multi-junction Cells

Cells made from multiple materials layers can have multiple bandgaps and will therefore respond to multiple light wavelengths, capturing and converting some of the energy that would otherwise be lost to relaxation.

For instance, if one had a cell with two bandgaps in it, one tuned to red light and the other to green, then the extra energy in green, cyan and blue light would be lost only to the bandgap of the green-sensitive material, while the energy of the red, yellow and orange would be lost only to the bandgap of the red-sensitive material. Following analysis similar to those performed for single-bandgap devices, it can be demonstrated that the perfect bandgaps for a two-gap device are at 1.1 eV and 1.8 eV.

Conveniently, light of a particular wavelength does not interact strongly with materials that are of bigger bandgap. This means that you can make a multi-junction cell by layering the different materials on top of each other, shortest wavelengths (biggest bandgap) on the "top" and increasing through the body of the cell. As the photons have to pass through the cell to reach the proper layer to be absorbed, transparent conductors need to be used to collect the electrons being generated at each layer.

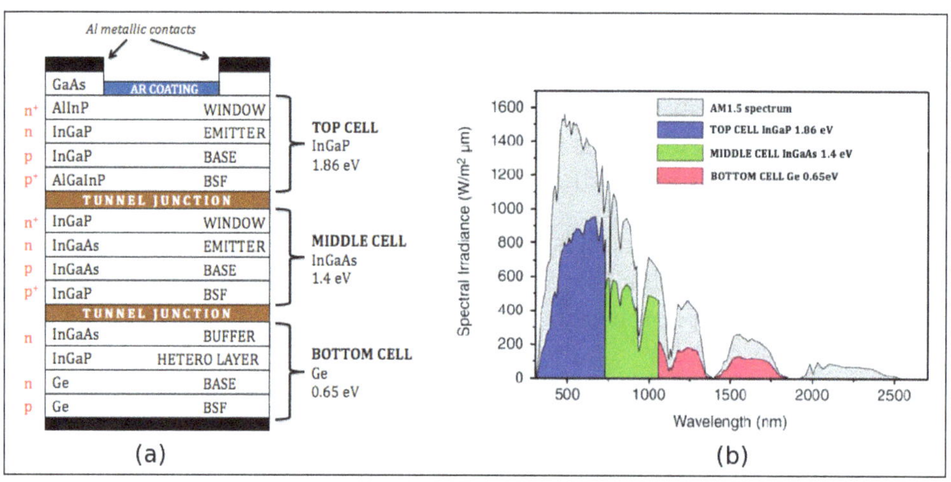

Figure. (a) The structure of an MJ solar cell. There are six important types of layers: pn junctions, back surface field (BSF) layers, window layers, tunnel junctions, anti-reflective coating and metallic contacts. (b) Graph of spectral irradiance E vs. wavelength λ over the AM1.5 solar spectrum, together with the maximum electricity conversion efficiency for every junction as a function of the wavelength.

Producing a tandem cell is not an easy task, largely due to the thinness of the materials and the difficulties extracting the current between the layers. The easy solution is to use two mechanically separate thin film solar cells and then wire them together separately outside the cell. This technique is widely used by amorphous silicon solar cells, Uni-Solar's products use three such layers to reach efficiencies around 9%. Lab examples using more exotic thin-film materials have demonstrated efficiencies over 30%.

The more difficult solution is the "monolithically integrated" cell, where the cell consists of a number of layers that are mechanically and electrically connected. These cells are much more difficult to produce because the electrical characteristics of each layer have to be carefully matched. In particular, the photocurrent generated in each layer needs to be matched, otherwise electrons will be absorbed between layers. This limits their construction to certain materials, best met by the III-V semiconductors.

Material Choice

The choice of materials for each sub-cell is determined by the requirements for lattice-matching, current-matching, and high performance opto-electronic properties.

For optimal growth and resulting crystal quality, the crystal lattice constant a of each material must be closely matched, resulting in lattice-matched devices. This constraint has been relaxed somewhat in recently developed metamorphic solar cells which contain a small degree of lattice mismatch. However, a greater degree of mismatch or other growth imperfections can lead to crystal defects causing a degradation in electronic properties.

Since each sub-cell is connected electrically in series, the same current flows through each junction. The materials are ordered with decreasing bandgaps, E_g, allowing sub-bandgap light ($hc/\lambda < e{\cdot}E_g$) to transmit to the lower sub-cells. Therefore, suitable bandgaps must be chosen such that the design spectrum will balance the current generation in each of the sub-cells, achieving current matching. Figure (b) plots spectral irradiance $E(\lambda)$, which is the source power density at a given wavelength λ. It is plotted together with the maximum conversion efficiency for every junction as a function of the wavelength, which is directly related to the number of photons available for conversion into photocurrent.

Finally, the layers must be electrically optimal for high performance. This necessitates usage of materials with strong absorption coefficients $\alpha(\lambda)$, high minority carrier lifetimes $\tau_{minority}$, and high mobilities μ.

The favorable values in the table below justify the choice of materials typically used for multi-junction solar cells: InGaP for the top sub-cell (E_g = 1.8 – 1.9 eV), InGaAs for the middle sub-cell (E_g = 1.4 eV), and Germanium for the bottom sub-cell (E_g = 0.67 eV). The use of Ge is mainly due to its lattice constant, robustness, low cost, abundance, and ease of production.

Because the different layers are closely lattice-matched, the fabrication of the device typically employs metal-organic chemical vapor deposition (MOCVD). This technique is preferable to the molecular beam epitaxy (MBE) because it ensures high crystal quality and large scale production.

Material	E_g, eV	a, nm	absorption (λ = 0.8 μm), 1/μm	μ_n, cm²/(V·s)	τ_p, μs	Hardness (Mohs)	α, μm/K	S, m/s
c-Si	1.12	0.5431	0.102	1400	1	7	2.6	0.1–60
InGaP	1.86	0.5451	2	500	–	5	5.3	50
GaAs	1.4	0.5653	0.9	8500	3	4–5	6	50

Ge	0.65	0.5657	3	3900	1000	6	7	1000
InGaAs	1.2	0.5868	30	1200	–	–	5.66	100–1000

Structural Elements

Metallic Contacts

The metallic contacts are low-resistivity electrodes that make contact with the semiconductor layers. They are often aluminum. This provides an electrical connection to a load or other parts of a solar cell array. They are usually on two sides of the cell. And are important to be on the back face so that shadowing on the lighting surface is reduced.

Anti-reflective Coating

Anti-reflective (AR) coating is generally composed of several layers in the case of MJ solar cells. The top AR layer has usually a NaOH surface texturation with several pyramids in order to increase the transmission coefficient *T*, the trapping of the light in the material (because photons cannot easily get out the MJ structure due to pyramids) and therefore, the path length of photons in the material. On the one hand, the thickness of each AR layer is chosen to get destructive interferences. Therefore, the reflection coefficient *R* decreases to 1%. In the case of two AR layers L_1 (the top layer, usually SiO_2) and L_2 (usually TiO_2), there must be $n_{L2} = n_{AlInP}^{1/2} \cdot n_{L1}$ to have the same amplitudes for reflected fields and $n_{L1}d_{L1} = 4\lambda_{min}, n_{L2}d_{L2} = \lambda_{min}/4$ to have opposite phase for reflected fields. On the other hand, the thickness of each AR layer is also chosen to minimize the reflectance at wavelengths for which the photocurrent is the lowest. Consequently, this maximizes J_{SC} by matching currents of the three subcells. As example, because the current generated by the bottom cell is greater than the currents generated by the other cells, the thickness of AR layers is adjusted so that the infrared (IR) transmission (which corresponds to the bottom cell) is degraded while the ultraviolet transmission (which corresponds to the top cell) is upgraded. Particularly, an AR coating is very important at low wavelengths because, without it, *T* would be strongly reduced to 70%.

Tunnel Junctions

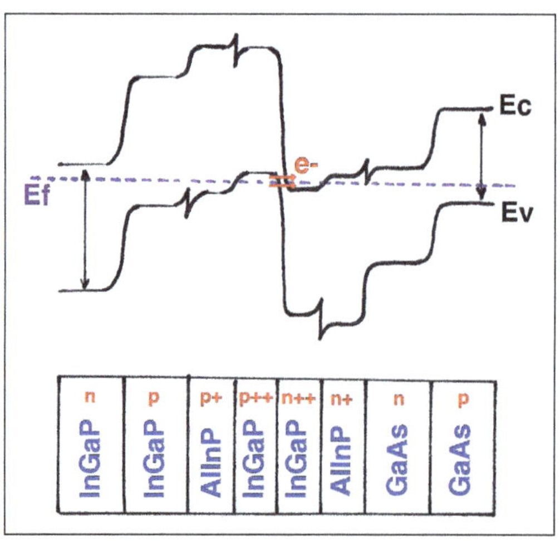

Layers and band diagram of the tunnel junction. Because the length of the depletion region is narrow and the band gap is high, electrons can tunnel.

The main goal of tunnel junctions is to provide a low electrical resistance and optically low-loss connection between two subcells. Without it, the p-doped region of the top cell would be directly connected with the n-doped region of the middle cell. Hence, a pn junction with opposite direction to the others would appear between the top cell and the middle cell. Consequently, the photovoltage would be lower than if there would be no parasitic diode. In order to decrease this effect, a tunnel junction is used. It is simply a wide band gap, highly doped diode. The high doping reduces the length of the depletion region because:

$$l_{depl} = \sqrt{\frac{2\grave{o}(\phi_0 - V)}{q}\frac{N_A + N_D}{N_A N_D}}$$

Hence, electrons can easily tunnel through the depletion region. The J-V characteristic of the tunnel junction is very important because it explains why tunnel junctions can be used to have a low electrical resistance connection between two pn junctions. Figure shows three different regions: the tunneling region, the negative differential resistance region and the thermal diffusion region. The region where electrons can tunnel through the barrier is called the tunneling region. There, the voltage must be low enough so that energy of some electrons who are tunneling is equal to energy states available on the other side of the barrier. Consequently, current density through the tunnel junction is high (with maximum value of , the peak current density) and the slope near the origin is therefore steep. Then, the resistance is extremely low and consequently, the voltage too. This is why tunnel junctions are ideal for connecting two pn junctions without having a voltage drop. When voltage is higher, electrons cannot cross the barrier because energy states are no longer available for electrons. Therefore, the current density decreases and the differential resistance is negative. The last region, called thermal diffusion region, corresponds to the J-V characteristic of the usual diode:

$$J = J_S(\exp(\frac{qV}{kT}) - 1)$$

In order to avoid the reduction of the MJ solar cell performances, tunnel junctions must be transparent to wavelengths absorbed by the next photovoltaic cell, the middle cell, i.e. $E_{gTunnel} > E_{gMiddleCell}$.

Window Layer and Back-surface Field

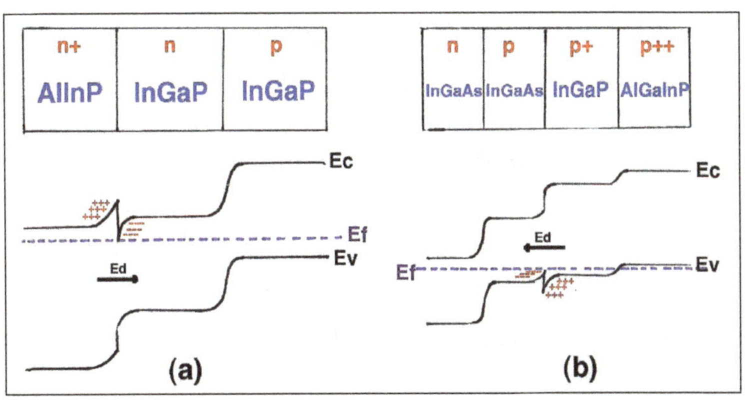

(a) Layers and band diagram of a window layer. The surface recombination is reduced.
(b) Layers and band diagram of a BSF layer. The scattering of carriers is reduced.

A window layer is used in order to reduce the surface recombination velocity *S*. Similarly, a back-surface field (BSF) layer reduces the scattering of carriers towards the tunnel junction. The structure of these two layers is the same: it is a heterojunction which catches electrons (holes). Indeed, despite the electric field E_d, these cannot jump above the barrier formed by the heterojunction because they don't have enough energy, as illustrated in figure. Hence, electrons (holes) cannot recombine with holes (electrons) and cannot diffuse through the barrier. By the way, window and BSF layers must be transparent to wavelengths absorbed by the next pn junction i.e. $E_{gWindow} > E_{gEmitter}$ and $E_{gBSF} > E_{gEmitter}$. Furthermore, the lattice constant must be close to the one of InGaP and the layer must be highly doped ($n \geq 10^{18}$ cm^{-3}).

J-V Characteristic

For maximum efficiency, each subcell should be operated at its optimal J-V parameters, which are not necessarily equal for each subcell. If they are different, the total current through the solar cell is the lowest of the three. By approximation, it results in the same relationship for the short-circuit current of the MJ solar cell: $J_{SC} = min\ (J_{SC1}, J_{SC2}, J_{SC3})$ where $J_{SCi}(\lambda)$ is the short-circuit current density at a given wavelength λ for the subcell *i*.

Because of the impossibility to obtain $J_{SC1}, J_{SC2}, J_{SC3}$ directly from the total J-V characteristic, the quantum efficiency $QE(\lambda)$ is utilized. It measures the ratio between the amount of electron-hole pairs created and the incident photons at a given wavelength λ. Let $\varphi_i(\lambda)$ be the photon flux of corresponding incident light in subcell *i*and$QE_i(\lambda)$ be the quantum efficiency of the subcell *i*. By definition, this equates to:

$$QE_i(\lambda) = \frac{J_{SCi}(\lambda)}{q\phi_i(\lambda)} \Rightarrow J_{SCi} = \int_0^{\lambda 2} q\phi_i(\lambda) QE_i(\lambda) d\lambda$$

The value of $QE_i(\lambda)$ is obtained by linking it with the absorption coefficient $\alpha(\lambda)$, i.e. the number of photons absorbed per unit of length by a material. If it is assumed that each photon absorbed by a subcell creates an electron/hole pair (which is a good approximation), this leads to:

$QE_i(\lambda) = 1 - e^{-\alpha(\lambda)d_i}$ where d_i is the thickness of the subcell *i* and $e^{-\alpha(\lambda)d_i}$ is the percentage of incident light which is not absorbed by the subcell *i*.

Similarly, because

$$V = \sum_{i=1}^{3} V_i,$$

the following approximation can be used:

$$V_{OC} = \sum_{i=1}^{3} V_{OCi}.$$

The values of V_{OCi} are then given by the J-V diode equation:

$$J_i = J_{0i}\left(e^{\frac{qV_i}{kT}} - 1\right) - J_{SCi} \Rightarrow V_{OCi} \approx \frac{kT}{q} \ln\left(\frac{J_{SCi}}{J_{0i}}\right)$$

Theoretical Limiting Efficiency

We can estimate the limiting efficiency of ideal infinite multi-junction solar cells using the graphical quantum-efficiency (QE) analysis invented by C. H. Henry. To fully take advantage of Henry's method, the unit of the AM1.5 spectral irradiance should be converted to that of photon flux (i.e., number of photons/m²/s). To do that, it is necessary to carry out an intermediate unit conversion from the power of electromagnetic radiation incident per unit area per photon energy to the photon flux per photon energy (i.e., from [W/m²/eV] to [number of photons/m2/s/eV]). For this intermediate unit conversion, the following points have to be considered: A photon has a distinct energy which is defined as follows:

- $E_{ph} = h \cdot f = h \cdot (c/\lambda)$

  where E_{ph} is photon energy, h is Planck's constant ($h = 6.626*10^{-34}$ [J·s]), c is speed of light ($c = 2.998*10^{8}$ [m/s]), f is frequency [1/s], and λ is wavelength [nm].

  Then the photon flux per photon energy, $dn_{ph}/dh\nu$, with respect to certain irradiance E [W/m²/eV] can be calculated as follows:

- $(\frac{dn_{ph}}{dh\nu} = \frac{E}{E_{ph}} = E/\{h \cdot (c/\lambda)\} = E[W/(m^2 \cdot eV)] \cdot \lambda \cdot (10^{-9}[m])/(1.998 \cdot 10^{-25}[J \cdot s \cdot m/s]) = E \cdot \lambda \cdot 5.03 \cdot 10^{15}$

  [(# of photons)/(m²·s·eV)]

As a result of this intermediate unit conversion, the AM1.5 spectral irradiance is given in unit of the photon flux per photon energy, [number of photons/m²/s/eV], as shown in Figure.

As a result of this intermediate unit conversion, the AM1.5 spectral irradiance is given in unit of the photon flux per photon energy, [number of photons/m²/s/eV], as shown in Figure.

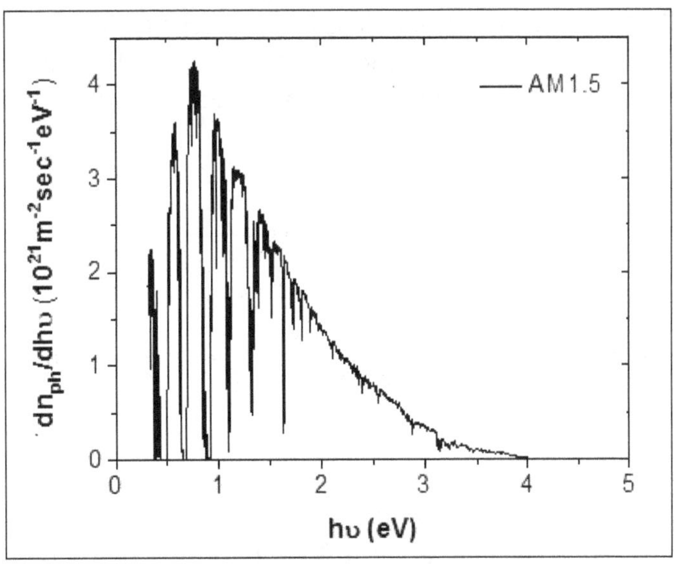

Photon flux per photon energy from standard solar energy spectrum (AM of 1.5).

Based on the above result from the intermediate unit conversion, we can derive the photon flux by numerically integrating the photon flux per photon energy with respect to photon energy. The numerically integrated photon flux is calculated using the Trapezoidal rule, as follows:

- $$n_{ph}(E_g) = \int_{E_g}^{\infty} \frac{dn_{ph}}{dh\nu} dh\nu = \sum_{i=E_g}^{\infty} (h\nu_{i+1} - h\nu_i) \frac{1}{2} [\frac{dn_{ph}}{dh\nu}(h\nu_{i+1}) + \frac{dn_{ph}}{dh\nu}(h\nu_i)]$$

As a result of this numerical integration, the AM1.5 spectral irradiance is given in unit of the photon flux, [number of photons/m2/s], as shown in figure below:

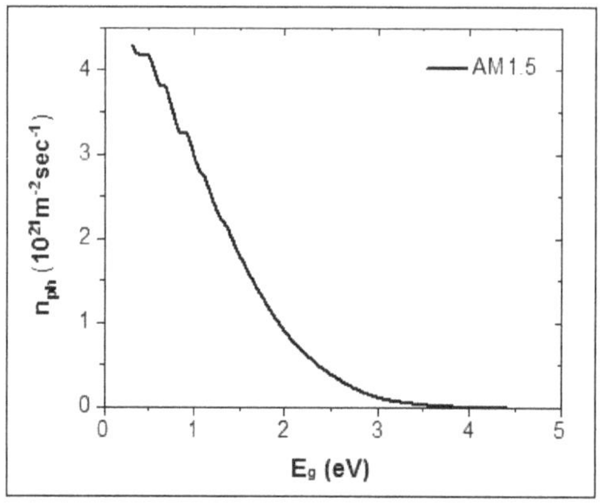

Photon flux from standard solar energy spectrum (AM of 1.5).

There are no photon flux data in the small photon energy range from 0 eV to 0.3096 eV because the standard (AM1.5) solar energy spectrum for hν < 0.31 eV are not available. Regardless of this data unavailability, however, the graphical QE analysis can be done using the only available data with a reasonable assumption that semiconductors are opaque for photon energies greater than their bandgap energy, but transparent for photon energies less than their bandgap energy. This assumption accounts for the first intrinsic loss in the efficiency of solar cells, which is caused by the inability of single-junction solar cells to properly match the broad solar energy spectrum. However, the current graphical QE analysis still cannot reflect the second intrinsic loss in the efficiency of solar cells, radiative recombination. To take the radiative recombination into account, we need to evaluate the radiative current density, J_{rad}, first. According to Shockley and Queisser method, J_{rad} can be approximated as follows.

- $$J_{rad} = A\exp(\frac{eV - E_g}{kT})$$

- $$A = \frac{2\pi \exp(n^2 + 1)E_g^2 kT}{h^3 c^2}$$

where Eg is in electron volts and n is evaluated to be 3.6, the value for GaAs. The incident absorbed thermal radiation Jth is given by Jrad with V = 0.

- $$J_{th} = A\exp\left(\frac{-E_g}{kT}\right)$$

The current density delivered to the load is the difference of the current densities due to absorbed solar and thermal radiation and the current density of radiation emitted from the top surface or absorbed in the substrate. Defining $J_{ph} = en_{ph}$, we have

- $J = J_{ph} + J_{th} - J_{rad}$

The second term, J_{th}, is negligible compared to J_{ph} for all semiconductors with E_g. ≥ 0.3 eV, as can be shown by evaluation of the above J_{th} equation. Thus, we will neglect this term to simplify the following discussion. Then we can express J as follows.

- $$J = en_{ph} - A\exp\left(\frac{eV - E_g}{kT}\right)$$

  The open-circuit voltage is found by setting J = 0.

- $$eV_{OC} = E_g - kT\ln\left(\frac{A}{en_{ph}}\right)$$

  The maximum power point (Jm, Vm) is found by stetting the derivative $\frac{dJV}{dV} = 0$. The familiar result of this calculation is:

- $$eV_m = eV_{OC} - kT\ln\left(1 + \frac{eV_m}{kT}\right)$$

- $$J_m = \frac{en_{ph}}{1 + kT / eV_m}$$

  Finally, the maximum work (Wm) done per absorbed photon, Wm is given by:

- $$W_m = \frac{J_m V_m}{n_{ph}} = \frac{eV_m}{1 + kT / eV_m} = eV_m - kT$$

  Combining the last three equations, we have

- $$W_m = E_g - kT[\ln(\frac{A}{en_{ph}}) + \ln(1 + \frac{eV_m}{kT}) + 1]$$

  Using the above equation, W_m (red line) is plotted in Figure for different values of E_g (or n_{ph}).

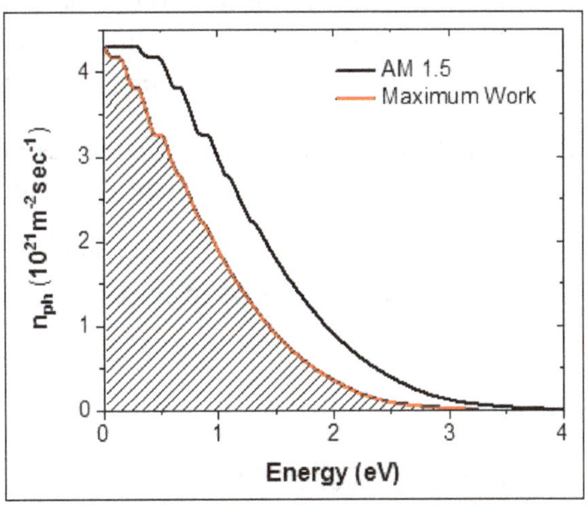

Maximum work by ideal infinite multi-junction solar cells under standard AM1.5 spectral irradiance.

Now, we can fully use Henry's graphical QE analysis, taking into account the two major intrinsic losses in the efficiency of solar cells. The two main intrinsic losses are radiative recombination, and the inability of single junction solar cells to properly match the broad solar energy spectrum. The shaded area under the red line represents the maximum work done by ideal infinite multi-junction solar cells. Hence, the limiting efficiency of ideal infinite multi-junction solar cells is evaluated to be 68.8% by comparing the shaded area defined by the red line with the total photon-flux area determined by the black line. (This is why this method is called "graphical" QE analysis.) Although this limiting efficiency value is consistent with the values published by Parrott and Vos in 1979: 64% and 68.2% respectively, there is a small gap between the estimated value in this report and literature values. This minor difference is most likely due to the different ways how to approximate the photon flux from 0 eV to 0.3096 eV. Here, we approximated the photon flux from 0 eV to 0.3096 eV as the same as the photon flux at 0.31 eV.

Materials

The majority of multi-junction cells that have been produced to date use three layers (although many tandem a-Si:H/mc-Si modules have been produced and are widely available). However, the triple junction cells require the use of semiconductors that can be tuned to specific frequencies, which has led to most of them being made of gallium arsenide (GaAs) compounds, often germanium for the bottom-, GaAs for the middle-, and $GaInP_2$ for the top-cell.

Gallium Arsenide Substrate

Dual junction cells can be made on Gallium arsenide wafers. Alloys of Indium gallium phosphide in the range $In_{.5}Ga_{.5}P$ through $In_{.53}Ga_{.47}P$ serve as the high band gap alloy. This alloy range provides for the ability to have band gaps in the range of 1.92eV to 1.87eV. The lower GaAs junction has a band gap of 1.42eV.

Germanium Substrate

Triple junction cells consisting of indium gallium phosphide (InGaP), gallium arsenide (GaAs) or indium gallium arsenide (InGaAs) and germanium (Ge) can be fabricated on germanium wafers. Early cells used straight gallium arsenide in the middle junction. Later cells have utilized $In_{0.015}Ga_{0.985}As$, due to the better lattice match to Ge, resulting in a lower defect density.

Due to the huge band gap difference between GaAs (1.42eV), and Ge (0.66eV), the current match is very poor, with the Ge junction operated significantly current limited.

Current efficiencies for commercial InGaP/GaAs/Ge cells approach 40% under concentrated sunlight. Lab cells (partly using additional junctions between the GaAs and Ge junction) have demonstrated efficiencies above 40%.

Indium Phosphide Substrate

Indium phosphide may be used as a substrate to fabricate cells with band gaps between 1.35eV and 0.74eV. Indium Phosphide has a band gap of 1.35eV. Indium gallium arsenide ($In_{0.53}Ga_{0.47}As$) is lattice matched to Indium Phosphide with a band gap of 0.74eV. A quaternary alloy of Indium gallium arsenide phosphide can be lattice matched for any band gap in between the two.

Indium phosphide-based cells have the potential to work in tandem with gallium arsenide cells. The two cells can be optically connected in series (with the InP cell below the GaAs cell), or in parallel through the use of spectra splitting using a Dichroic filter.

Indium Gallium Nitride Substrate

Indium gallium nitride (InGaN) is a semiconductor material made of a mix of gallium nitride (GaN) and indium nitride (InN). It is a ternary group III/V direct bandgap semiconductor. Its bandgap can be tuned by varying the amount of indium in the alloy from 0.7 eV to 3.4 eV, thus making it an ideal material for solar cells. However, its conversion efficiencies because of technological factors unrelated to bandgap are still not high enough to be competitive in the market.

Performance Improvements

Structure

Many MJ photovoltaic cells use III-V semiconductor materials. GaAsSb-based heterojunction tunnel diodes, instead of conventional InGaP highly doped tunnel diodes have a lower tunneling distance. Indeed, in the heterostructure formed by GaAsSb and InGaAs, the valence band of GaAsSb is higher than the valence band of the adjoining p-doped layer. Consequently, the tunneling distance d_{tunnel} is reduced and so the tunneling current, which exponentially depends of d_{tunnel}, is increased. Hence, the voltage is lower than that of the InGaP tunnel junction. GaAsSb heterojunction tunnel diodes offer other advantages. The same current can be achieved by using a lower doping. Secondly, because the lattice constant is larger for GaAsSb than Ge, one can use a wider range of materials for the bottom cell because more materials are lattice-matched to GaAsSb than to Ge.

Chemical components can be added to some layers. Adding about one percent of Indium in each layer better matches lattice constants of the different layers. Without it, there is about 0.08 percent of mismatching between layers, which inhibits performance. Adding aluminium to the top cell increases its band gap to 1.96 eV, covering a larger part of the solar spectrum and obtain a higher open-circuit voltage V_{OC}.

The theoretical efficiency of MJ solar cells is 86.8% for an infinite number of pn junctions, implying that more junctions increase efficiency. The maximum theoretical efficiency is 37, 50, 56, 72% for 1, 2, 3, 36 pn junctions, respectively, with the number of junctions increasing exponentially to achieve equal efficiency increments. The exponential relationship implies that as the cell approaches the limit of efficiency, the increase cost and complexity grow rapidly. Decreasing the thickness of the top cell increases the transmission coefficient T.

Finally, an InGaP hetero-layer between the p-Ge layer and the InGaAs layer can be added in order to create automatically the n-Ge layer by scattering during MOCVD growth and increase significantly the quantum efficiency $QE(\lambda)$ of the bottom cell. InGaP is advantageous because of its high scattering coefficient and low solubility in Ge.

Spectral Variations

Solar spectrum at the Earth surface changes constantly depending on the weather and sun position. This results in the variation of $\varphi(\lambda)$, $QE(\lambda)$, $\alpha(\lambda)$ and thus the short-circuit currents J_{SCi}.

As a result, the current densities J_i are not necessarily matched and the total current becomes lower. These variations can be quantified using the average photon energy (APE) which is the ratio between the spectral irradiance G(λ) (the power density of the light source in a specific wavelength λ) and the total photon flux density. It can be shown that a high (low) value for APE means low (high) wavelengths spectral conditions and higher (lower) efficiencies. Thus APE is a good indicator for quantifying the effects of the solar spectrum variations on performances and has the added advantage of being independent of the device structure and the absorption profile of the device.

Use of Light Concentrators

Light concentrators increase efficiencies and reduce the cost/efficiency ratio. The three types of light concentrators in use are refractive lenses like Fresnel lenses, reflective dishes (parabolic or cassegraine), and light guide optics. Thanks to these devices, light arriving on a large surface can be concentrated on a smaller cell. The intensity concentration ratio (or "suns") is the average intensity of the focused light divided by 1 kW/m^2 (reasonable value related to solar constant). If its value is X then the MJ current becomes X higher under concentrated illumination.

Using concentrations on the order of 500 to 1000, meaning that a 1 cm^2 cell can use the light collected from 0.1 m^2 (as 1 m^2 equal 10000 cm^2), produces the highest efficiencies seen to date. Three-layer cells are fundamentally limited to 63%, but existing commercial prototypes have already demonstrated over 40%. These cells capture about 2/3 of their theoretical maximum performance, so assuming the same is true for a non-concentrated version of the same design, one might expect a three-layer cell of 30% efficiency. This is not enough of an advantage over traditional silicon designs to make up for their extra production costs. For this reason, almost all multi-junction cell research for terrestrial use is dedicated to concentrator systems, normally using mirrors or fresnel lenses.

Using a concentrator also has the added benefit that the number of cells needed to cover a given amount of ground area is greatly reduced. A conventional system covering 1 m^2 would require 625 16 cm^2 cells, but for a concentrator system only a single cell is needed, along with a concentrator. The argument for concentrated Multi-junction cells has been that the high cost of the cells themselves would be more than offset by the reduction in total number of cells. However, the downside of the concentrator approach is that efficiency drops off very quickly under lower lighting conditions. In order to maximize its advantage over traditional cells and thus be cost competitive, the concentrator system has to track the sun as it moves to keep the light focused on the cell and maintain maximum efficiency as long as possible. This requires a solar tracker system, which increases yield, but also cost.

Fabrication

As of 2014 multi-junction cells were expensive to produce, using techniques similar to semiconductor device fabrication, usually metalorganic vapour phase epitaxy but on "chip" sizes on the order of centimeters.

A new technique was announced that year that allowed such cells to use a substrate of glass or steel, lower-cost vapors in reduced quantities that was claimed to offer costs competitive with conventional silicon cells.

Applications

As of 2010, the cost of MJ solar cells was too high to allow use outside of specialized applications. The high cost is mainly due to the complex structure and the high price of materials. Nevertheless, with light concentrators under illumination of at least 400 suns, MJ solar panels become practical.

As less expensive multi-junction materials become available other applications involve bandgap engineering for microclimates with varied atmospheric conditions.

MJ cells are currently being utilized in the Mars rover missions.

The environment in space is quite different. Because there is no atmosphere, the solar spectrum is different (AM0). The cells have a poor current match due to a greater photon flux of photons above 1.87eV vs. those between 1.87eV and 1.42eV. This results in too little current in the GaAs junction, and hampers the overall efficiency since the InGaP junction operates below MPP current and the GaAs junction operates above MPP current. To improve current match, the InGaP layer is intentionally thinned to allow additional photons to penetrate to the lower GaAs layer.

In terrestrial concentrating applications, the scatter of blue light by the atmosphere reduces the photon flux above 1.87eV, better balancing the junction currents. Radiation particles that are no longer filtered can damage the cell. There are two kinds of damage: ionisation and atomic displacement. Still, MJ cells offer higher radiation resistance, higher efficiency and a lower temperature coefficient.

Organic Solar Cell

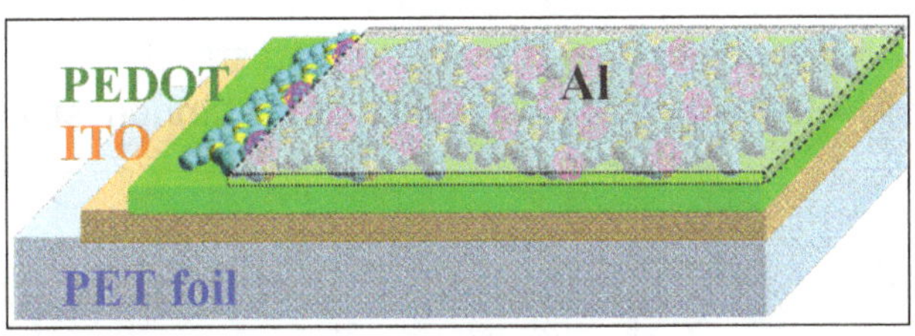

Schematic of plastic solar cells. PET – polyethylene terephthalate, ITO – indium tin oxide, PEDOT:PSS – poly(3,4-ethylenedioxythiophene), active layer (usually a polymer:fullerene blend), Al – aluminium.

An organic solar cell (OSC) or plastic solar cell is a type of photovoltaic that uses organic electronics, a branch of electronics that deals with conductive organic polymers or small organic molecules, for light absorption and charge transport to produce electricity from sunlight by the photovoltaic effect. Most organic photovoltaic cells are polymer solar cells.

The molecules used in organic solar cells are solution-processable at high throughput and are cheap, resulting in low production costs to fabricate a large volume. Combined with the flexibility of organic molecules, organic solar cells are potentially cost-effective for photovoltaic applications. Molecular engineering (e.g. changing the length and functional group of polymers) can change the band gap, allowing for electronic tunability. The optical absorption coefficient of organic molecules is high, so a large amount of light can be absorbed with a small amount of materials, usually on the order of hundreds of nanometers. The main disadvantages associated with organic photovoltaic cells are low efficiency, low stability and low strength compared to inorganic photovoltaic cells such as silicon solar cells.

Organic Photovoltaic manufactured by the company, Solarmer.

Compared to silicon-based devices, polymer solar cells are lightweight (which is important for small autonomous sensors), potentially disposable and inexpensive to fabricate (sometimes using printed electronics), flexible, customizable on the molecular level and potentially have less adverse environmental impact. Polymer solar cells also have the potential to exhibit transparency, suggesting applications in windows, walls, flexible electronics, etc. An example device is shown in Fig. The disadvantages of polymer solar cells are also serious: they offer about 1/3 of the efficiency of hard materials, and experience substantial photochemical degradation.

Polymer solar cells inefficiency and stability problems, combined with their promise of low costs and increased efficiency made them a popular field in solar cell research. As of 2015, polymer solar cells were able to achieve over 10% efficiency via a tandem structure. In 2018, a record breaking efficiency for organic photovoltaics of 17.3% was reached via tandem structure.

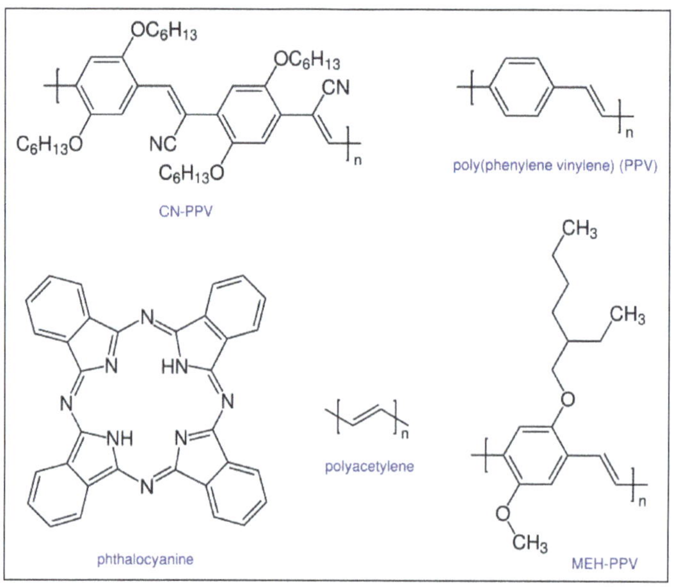

Examples of organic photovoltaic materials.

A photovoltaic cell is a specialized semiconductor diode that converts light into direct current (DC) electricity. Depending on the band gap of the light-absorbing material, photovoltaic cells can also convert low-energy, infrared (IR) or high-energy, ultraviolet (UV) photons into DC electricity.

A common characteristic of both the small molecules and polymers used as the light-absorbing material in photovoltaics is that they all have large conjugated systems. A conjugated system is formed where carbon atoms covalently bond with alternating single and double bonds. These hydrocarbons' electrons pz orbitals delocalize and form a delocalized bonding π orbital with a π^* antibonding orbital. The delocalized π orbital is the highest occupied molecular orbital (HOMO), and the π^* orbital is the lowest unoccupied molecular orbital (LUMO). In organic semiconductor physics, the HOMO takes the role of the valence band while the LUMO serves as the conduction band. The energy separation between the HOMO and LUMO energy levels is considered the band gap of organic electronic materials and is typically in the range of 1–4 eV.

All light with energy greater than the band gap of the material can be absorbed, though there is a trade-off to reducing the band gap as photons absorbed with energies higher than the band gap will thermally give off its excess energy, resulting in lower voltages and power conversion efficiencies. When these materials absorb a photon, an excited state is created and confined to a molecule or a region of a polymer chain. The excited state can be regarded as an exciton, or an electron-hole pair bound together by electrostatic interactions. In photovoltaic cells, excitons are broken up into free electron-hole pairs by effective fields. The effective fields are set up by creating a heterojunction between two dissimilar materials. In organic photovoltaics, effective fields break up excitons by causing the electron to fall from the conduction band of the absorber to the conduction band of the acceptor molecule. It is necessary that the acceptor material has a conduction band edge that is lower than that of the absorber material.

Polymer solar cells usually consist of an electron- or hole-blocking layer on top of an indium tin oxide (ITO) conductive glass followed by electron donor and an electron acceptor (in the case of bulk heterojunction solar cells), a hole or electron blocking layer, and metal electrode on top. The nature and order of the blocking layers – as well as the nature of the metal electrode – depends on whether the cell follows a regular or an inverted device architecture. In an inverted cell, the electric charges exit the device in the opposite direction as in a normal device because the positive and negative electrodes are reversed. Inverted cells can utilize cathodes out of a more suitable material; inverted OPVs enjoy longer lifetimes than regularly structured OPVs, but they typically don't reach efficiencies as high as regular OPVs.

In bulk heterojunction polymer solar cells, light generates excitons. Subsequent charge separation in the interface between an electron donor and acceptor blend within the device's active layer. These charges then transport to the device's electrodes where the charges flow outside the cell, perform work and then re-enter the device on the opposite side. The cell's efficiency is limited by several factors, especially non-geminate recombination. Hole mobility leads to faster conduction across the active layer.

Organic photovoltaics are made of electron donor and electron acceptor materials rather than semiconductor p-n junctions. The molecules forming the electron donor region of organic PV cells, where exciton electron-hole pairs are generated, are generally conjugated polymers possessing delocalized π electrons that result from carbon p orbital hybridization. These π electrons can be excited by light in or near the visible part of the spectrum from the molecule's highest occupied molecular orbital (HOMO) to the lowest unoccupied molecular orbital (LUMO), denoted by a π -π^* transition. The energy bandgap between these orbitals determines which wavelength(s) of light can be absorbed.

Unlike in an inorganic crystalline PV cell material, with its band structure and delocalized electrons, excitons in organic photovoltaics are strongly bound with an energy between 0.1 and 1.4 eV. This strong binding occurs because electronic wave functions in organic molecules are more localized, and electrostatic attraction can thus keep the electron and hole together as an exciton. The electron and hole can be dissociated by providing an interface across which the chemical potential of electrons decreases. The material that absorbs the photon is the donor, and the material acquiring the electron is called the acceptor. In Fig., the polymer chain is the donor and the fullerene is the acceptor. Even after dissociation, the electron and hole may still be joined as a "geminate pair", and an electric field is then required to separate them. The electron and hole must be collected at contacts. If charge carrier mobility is insufficient, the carriers will not reach the contacts, and instead recombine at trap sites or remain in the device as undesirable space charges that oppose the flow of new carriers. The latter problem can occur if electron and hole mobilities are not matched. In that case, space-charge limited photocurrent (SCLP) hampers device performance.

Organic photovoltaics can be fabricated with an active polymer and a fullerene-based electron acceptor. Illumination of this system by visible light leads to electron transfer from the polymer to a fullerene molecule. As a result, the formation of a photoinduced quasiparticle, or polaron (P^+), occurs on the polymer chain and the fullerene becomes a radical anion (C^-_{60}). Polarons are highly mobile and can diffuse away.

Junction Types

The simplest organic PV device features a planar heterojunction. A film of organic active material (polymer or small molecule), of electron donor or electron acceptor type is sandwiched between contacts. Excitons created in the active material may diffuse before recombining and separate, hole and electron diffusing to its specific collecting electrode. Because charge carriers have diffusion lengths of just 3–10 nm in typical amorphous organic semiconductors, planar cells must be thin, but the thin cells absorb light less well. Bulk heterojunctions (BHJs) address this shortcoming. In a BHJ, a blend of electron donor and acceptor materials is cast as a mixture, which then phase-separates. Regions of each material in the device are separated by only several nanometers, a distance suited for carrier diffusion. BHJs require sensitive control over materials morphology on the nanoscale. Important variables include materials, solvents and the donor-acceptor weight ratio.

The next logical step beyond BHJs are ordered nanomaterials for solar cells, or ordered heterojunctions (OHJs). OHJs minimize the variability associated with BHJs. OHJs are generally hybrids of ordered inorganic materials and organic active regions. For example, a photovoltaic polymer can be deposited into pores in a ceramic such as TiO_2. Since holes still must diffuse the length of the pore through the polymer to a contact, OHJs suffer similar thickness limitations. Mitigating the hole mobility bottleneck is key to further enhancing device performance of OHJ's.

Single Layer

Single layer organic photovoltaic cells are the simplest form. These cells are made by sandwiching a layer of organic electronic materials between two metallic conductors, typically a layer of indium tin oxide (ITO) with high work function and a layer of low work function metal such as Aluminum, Magnesium or Calcium. The basic structure of such a cell is illustrated in Figure.

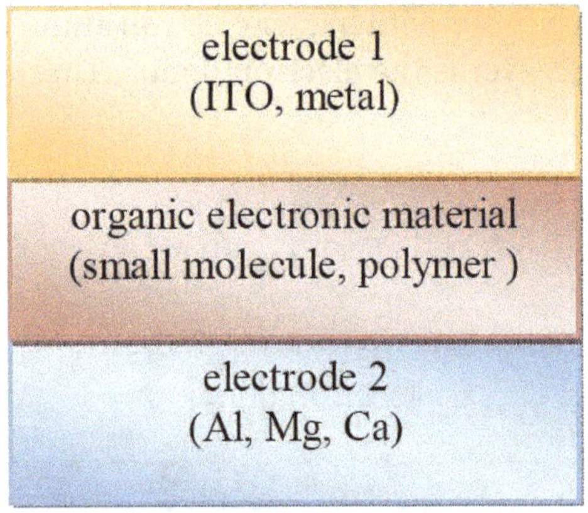

Sketch of a single layer organic photovoltaic cell.

The difference of work function between the two conductors sets up an electric field in the organic layer. When the organic layer absorbs light, electrons will be excited to the LUMO and leave holes in the HOMO, thereby forming excitons. The potential created by the different work functions helps to split the exciton pairs, pulling electrons to the positive electrode (an electrical conductor used to make contact with a non-metallic part of a circuit) and holes to the negative electrode.

Examples:

In 1958 the photovoltaic effect or the creation of voltage of a cell based on magnesium phthalocyanine (MgPc)—a macrocyclic compound having an alternating nitrogen atom-carbon atom ring structure—was discovered to have a photovoltage of 200 mV. An Al/MgPc/Ag cell obtained photovoltaic efficiency of 0.01% under illumination at 690 nm.

Conjugated polymers were also used in this type of photovoltaic cell. One device used polyacetylene as the organic layer, with Al and graphite, producing an open circuit voltage of 0.3 V and a charge collection efficiency of 0.3%. An Al/poly(3-nethyl-thiophene)/Pt cell had an external quantum yield of 0.17%, an open circuit voltage of 0.4 V and a fill factor of 0.3. An ITO/PPV/Al cell showed an open circuit voltage of 1 V and a power conversion efficiency of 0.1% under white-light illumination.

Issues

Single layer organic solar cells do not work well. They have low quantum efficiencies (<1%) and low power conversion efficiencies (<0.1%). A major problem with them is that the electric field resulting from the difference between the two conductive electrodes is seldom sufficient to split the excitons. Often the electrons recombine with the holes without reaching the electrode.

Bilayer

Bilayer cells contain two layers in between the conductive electrodes. The two layers have different electron affinity and ionization energies, therefore electrostatic forces are generated at the interface between the two layers. Light must create excitons in this small charged region for an efficient charge separation and collecting. The materials are chosen to make the differences large enough that these local electric fields are strong, which splits excitons much more efficiently than single layer photovoltaic cells. The layer with higher electron affinity and ionization potential is the electron acceptor, and the other layer is the electron donor. This structure is also called a planar donor-acceptor heterojunction.

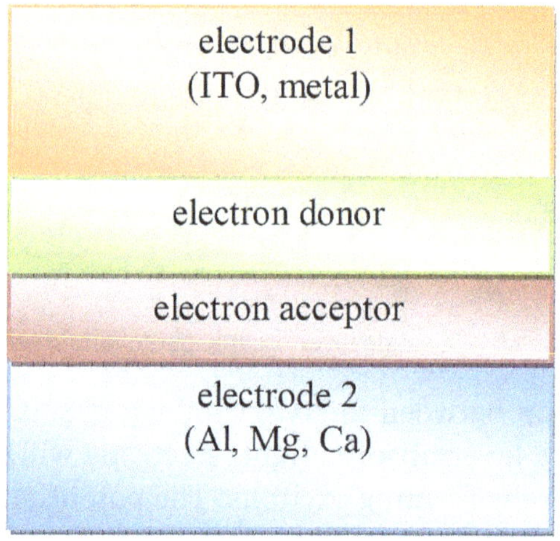

Sketch of a multilayer organic photovoltaic cell.

Examples:

C_{60} has high electron affinity, making it a good acceptor. A C_{60}/MEH-PPV double layer cell had a relatively high fill factor of 0.48 and a power conversion efficiency of 0.04% under monochromatic illumination. PPV/C_{60} cells displayed a monochromatic external quantum efficiency of 9%, a power conversion efficiency of 1% and a fill factor of 0.48.

Perylene derivatives display high electron affinity and chemical stability. A layer of copper phthalocyanine (CuPc) as electron donor and perylene tetracarboxylic derivative as electron acceptor, fabricating a cell with a fill factor as high as 0.65 and a power conversion efficiency of 1% under simulated AM2 illumination. Halls et al. fabricated a cell with a layer of bis (phenethylimido) perylene over a layer of PPV as the electron donor. This cell had peak external quantum efficiency of 6% and power conversion efficiency of 1% under monochromatic illumination, and a fill factor of up to 0.6.

Issues

The diffusion length of excitons in organic electronic materials is typically on the order of 10 nm. In order for most excitons to diffuse to the interface of layers and split into carriers, the layer thickness should be in the same range as the diffusion length. However, a polymer layer typically needs a thickness of at least 100 nm to absorb enough light. At such a large thickness, only a small fraction of the excitons can reach the heterojunction interface.

Discrete Heterojunction

A three-layer (two acceptor and one donor) fullerene-free stack achieved a conversion efficiency of 8.4%. The implementation produced high open-circuit voltages and absorption in the visible spectra and high short-circuit currents. Quantum efficiency was above 75% between 400 nm and 720 nm wavelengths, with an open-circuit voltage around 1 V.

Bulk Heterojunction

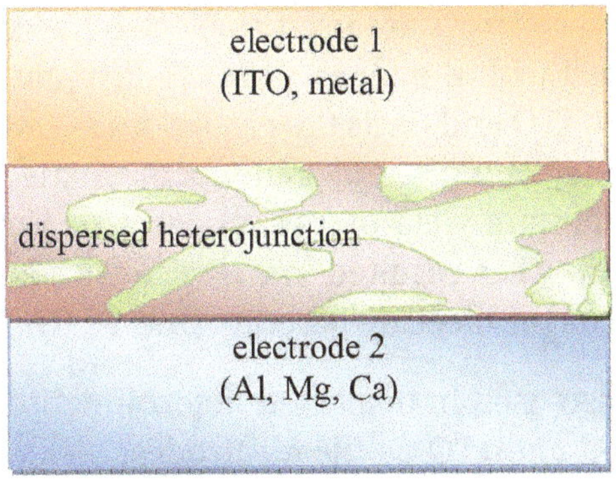

Sketch of a dispersed junction photovoltaic cell.

Bulk heterojunctions have an absorption layer consisting of a nanoscale blend of donor and acceptor materials. The domain sizes of this blend are on the order of nanometers, allowing for excitons with short lifetimes to reach an interface and dissociate due to the large donor-acceptor interfacial area. However, efficient bulk heterojunctions need to maintain large enough domain sizes to form a percolating network that allows the donor materials to reach the hole transporting electrode and the acceptor materials to reach the electron transporting electrode (Electrode 2). Without this percolating network, charges might be trapped in a donor or acceptor rich domain and undergo recombination. Bulk heterojunctions have an advantage over layered photoactive structures because they can be made thick enough for effective photon absorption without the difficult processing involved in orienting a layered structure while retaining similar level of performances.

Bulk heterojunctions are most commonly created by forming a solution containing the two components, casting (e.g. drop casting and spin coating) and then allowing the two phases to separate, usually with the assistance of an annealing step. The two components will self-assemble into an interpenetrating network connecting the two electrodes. They are normally composed of a conjugated molecule based donor and fullerene based acceptor. The nanostructural morphology of bulk heterojunctions tends to be difficult to control, but is critical to photovoltaic performance.

After the capture of a photon, electrons move to the acceptor domains, then are carried through the device and collected by one electrode, and holes move in the opposite direction and collected at the other side. If the dispersion of the two materials is too fine, it will result in poor charge transfer through the layer.

Most bulk heterojunction cells use two components, although three-component cells have been explored. The third component, a secondary p-type donor polymer, acts to absorb light in a different region of the solar spectrum. This in theory increases the amount of absorbed light. These ternary cells operate through one of three distinct mechanisms: charge transfer, energy transfer or parallel-linkage.

In charge transfer, both donors contribute directly to the generation of free charge carriers. Holes pass through only one donor domain before collection at the anode. In energy transfer, only one donor contributes to the production of holes. The second donor acts solely to absorb light, transferring extra energy to the first donor material. In parallel linkage, both donors produce excitons independently, which then migrate to their respective donor/acceptor interfaces and dissociate.

Examples:

Fullerenes such as C_{60} and its derivatives are used as electron acceptor materials in bulk heterojunction photovoltaic cells. A cell with the blend of MEH-PPV and a methano-functionalized C_{60} derivative as the heterojunction, ITO and Ca as the electrodes showed a quantum efficiency of 29% and a power conversion efficiency of 2.9% under monochromatic illumination. Replacing MEH-PPV with P3HT produced a quantum yield of 45% under a 10 V reverse bias. Further advances in modifying the electron acceptor has resulted in a device with a power conversion efficiency of 10.61% with a blend of $PC_{71}BM$ as the electron acceptor and PTB7-Th as the electron donor.

Polymer/polymer blends are also used in dispersed heterojunction photovoltaic cells. A blend of CN-PPV and MEH-PPV with Al and ITO as the electrodes, yielded peak monochromatic power conversion efficiency of 1% and fill factor of 0.38.

Dye sensitized photovoltaic cells can also be considered important examples of this type.

Issues

Fullerenes such as $PC_{71}BM$ are often the electron acceptor materials found in high performing bulk heterojunction solar cells. However, these electron acceptor materials very weakly absorb visible light, decreasing the volume fraction occupied by the strongly absorbing electron donor material. Furthermore, fullerenes have poor electronic tunability, resulting in restrictions placed on the development of conjugated systems with more appealing electronic structures for higher voltages. Recent research has been done on trying to replace these fullerenes with organic molecules that can be electronically tuned and contribute to light absorption.

Graded Heterojunction

The electron donor and acceptor are mixed in such a way that the gradient is gradual. This architecture combines the short electron travel distance in the dispersed heterojunction with the advantage of the charge gradient of the bilayer technology.

Examples:

A cell with a blend of CuPc and C_{60} showed a quantum efficiency of 50% and a power conversion efficiency of 2.1% using 100 mW/cm^2 simulated AM1.5G solar illumination for a graded heterojunction.

Continuous Junction

Similar to the graded heterojunction the continuous junction concept aims at realizing a gradual transition from an electron donor to an electron acceptor. However, the acceptor material is prepared directly from the donor polymer in a post-polymerization modification step.

Production

Since its active layer largely determines device efficiency, this component's morphology received much attention.

If one material is more soluble in the solvent than the other, it will deposit first on top of the substrate, causing a concentration gradient through the film. This has been demonstrated for poly-3-hexyl thiophene (P3HT), phenyl-C_{61}-butyric acid methyl ester (PCBM) devices where the PCBM tends to accumulate towards the device's bottom upon spin coating from ODCB solutions. This effect is seen because the more soluble component tends to migrate towards the "solvent rich" phase during the coating procedure, accumulating the more soluble component towards the film's bottom, where the solvent remains longer. The thickness of the generated film affects the phases segregation because the dynamics of crystallization and precipitation are different for more concentrated solutions or faster evaporation rates (needed to build thicker devices). Crystalline P3HT enrichment closer to the hole-collecting electrode can only be achieved for relatively thin (100 nm) P3HT/PCBM layers.

The gradients in the initial morphology are then mainly generated by the solvent evaporation rate and the differences in solubility between the donor and acceptor inside the blend. This dependence on solubility has been clearly demonstrated using fullerene derivatives and P3HT. When using solvents which evaporate at a slower rate (as chlorobenzene (CB) or dichlorobenzene (DCB)) you can get larger degrees of vertical separation or aggregation while solvents that evaporate quicker produce a much less effective vertical separation. Larger solubility gradients should lead to more effective vertical separation while smaller gradients should lead to more homogeneous films. These two effects were verified on P3HT:PCBM solar cells.

The solvent evaporation speed as well as posterior solvent vapor or thermal annealing procedures were also studied. Blends such as P3HT:PCBM seem to benefit from thermal annealing procedures, while others, such as PTB7:PCBM, seem to show no benefit. In P3HT the benefit seems to come from an increase of crystallinity of the P3HT phase which is generated through an expulsion of PCBM molecules from within these domains. This has been demonstrated through studies of PCBM miscibility in P3HT as well as domain composition changes as a function of annealing times.

The above hypothesis based on miscibility does not fully explain the efficiency of the devices as solely pure amorphous phases of either donor or acceptor materials never exist within bulk heterojunction devices. A 2010 paper suggested that current models that assume pure phases and discrete interfaces might fail given the absence of pure amorphous regions. Since current models assume phase separation at interfaces without any consideration for phase purity, the models might need to be changed.

The thermal annealing procedure varies depending on precisely when it is applied. Since vertical species migration is partly determined by the surface tension between the active layer and either air or another layer, annealing before or after the deposition of additional layers (most often the metal cathode) affects the result. In the case of P3HT:PCBM solar cells vertical migration is improved when cells are annealed after the deposition of the metal cathode.

Donor or acceptor accumulation next to the adjacent layers might be beneficial as these accumulations can lead to hole or electron blocking effects which might benefit device performance. In 2009 the difference in vertical distribution on P3HT:PCBM solar cells was shown to cause problems with electron mobility which ends up with the yielding of very poor device efficiencies. Simple changes to device architecture – spin coating a thin layer of PCBM on top of the P3HT – greatly enhance cell reproducibility, by providing reproducible vertical separation between device components.

Since higher contact between the PCBM and the cathode is required for better efficiencies, this largely increases device reproducibility.

According to neutron scattering analysis, P3HT:PCBM blends have been described as "rivers" (P3HT regions) interrupted by "streams" (PCBM regions).

Solvent Effects

Conditions for spin coating and evaporation affect device efficiency. Solvent and additives influence donor-acceptor morphology. Additives slow down evaporation, leading to more crystalline polymers and thus improved hole conductivities and efficiencies. Typical additives include 1,8-octanedithiol, ortho-dichlorobenzene, 1,8-diiodooctane (DIO), and nitrobenzene. The DIO effect was attributed to the selective solubilization of PCBM components, modifies fundamentally the average hopping distance of electrons, and thus improves electron mobility. Additives can also lead to big increases in efficiency for polymers. For HXS-1/PCBM solar cells, the effect was correlated with charge generation, transport and shelf-stability. Other polymers such as PTTBO also benefit significantly from DIO, achieving PCE values of more than 5% from around 3.7% without the additive.

Polymer Solar Cells fabricated from chloronaphthalene (CN) as a co-solvent enjoy a higher efficiency than those fabricated from the more conventional pure chlorobenzene solution. This is because the donor-acceptor morphology changes, which reduces the phase separation between donor polymer and fullerene. As a result, this translates into high hole mobilities. Without co-solvents, large domains of fullerene form, decreasing photovoltaic performance of the cell due to polymer aggregation in solution. This morphology originates from the liquid-liquid phase separation during drying; solve evaporation causes the mixture to enter into the spinodal region, in which there are significant thermal fluctuations. Large domains prevent electrons from being collected efficiently (decreasing PCE).

Small differences in polymer structure can also lead to significant changes in crystal packing that inevitably affect device morphology. PCPDTBT differs from PSBTBT caused by the difference in bridging atom between the two polymers (C vs. Si), which implies that better morphologies are achievable with PCPDTBT:PCBM solar cells containing additives as opposed to the Si system which achieves good morphologies without help from additional substances.

Self-assembled Cells

Supramolecular chemistry was investigated, using donor and acceptor molecules that assemble upon spin casting and heating. Most supramolecular assemblies employ small molecules. Donor and acceptor domains in a tubular structure appear ideal for organic solar cells.

Diblock polymers containing fullerene yield stable organic solar cells upon thermal annealing. Solar cells with pre-designed morphologies resulted when appropriate supramolecular interactions are introduced.

Progress on BCPs containing polythiophene derivatives yield solar cells that assemble into well defined networks. This system exhibits a PCE of 2.04%. Hydrogen bonding guides the morphology.

Device efficiency based on co-polymer approaches have yet to cross the 2% barrier, whereas bulk-heterojunction devices exhibit efficiencies >7% in single junction configurations.

Fullerene-grafted rod-coil block copolymers have been used to study domain organization.

Supramolecular approaches to organic solar cells provide understanding about the macromolecular forces that drive domain separation.

Infrared Polymer Cells

Infrared cells preferentially absorb light in the infrared range rather than visible wavelengths. As of 2012, such cells can be made nearly 70% transparent to visible light. The cells allegedly can be made in high volume at low cost using solution processing. Infrared polymer cells can be used as add-on components of portable electronics, smart windows, and building-integrated photovoltaics.The cells employ silver nanowire/titanium dioxide composite films as the top electrode, replacing conventional opaque metal electrodes. With this combination, 4% power-conversion efficiency was achieved.

Near-infrared Polymer solar cells based on a copolymer of naphthodithiophene diimide and bithiophene (PNDTI-BT-DT) are also being fabricated in combination with PTB7 as an electron donor. Both PNDTI-BT-DT and PTB7 formed a crystalline structure in the blend films similar to in the pristine films, leading to the efficient charge generation contributed from both polymers.

Much research has been focused on developing a transparent top electrode for PSCs. However, a 2017 study explored optimizing the active layer of semi-transparent PSCs. The researchers proposed a semi-transparent PSC with enhanced efficiency that utilizes both narrow bandgap polymer donor, PTB7-Th, and non-fullerene acceptor, IHIC. The results of this study showed that the proposed PSC exhibited high transmittance and absorption in the infrared spectrum but low absorption in the visible spectrum. This cell showed to be relatively stable and have a maximum PCE of 9.77%, which, as of 2017, is the highest reported PCE value.

Typical Current-voltage Behavior and Power Conversion Efficiency

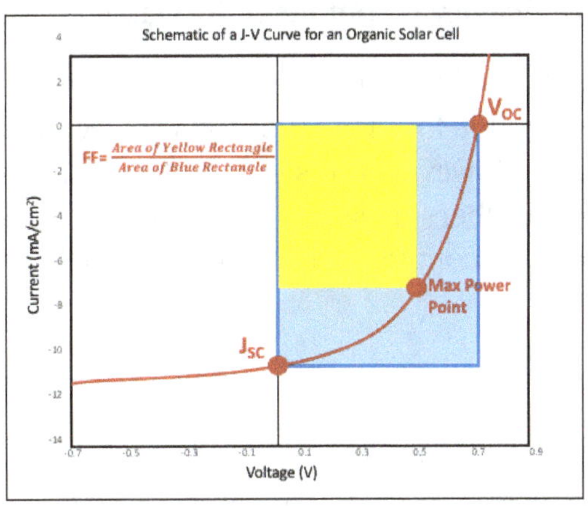

Organic photovoltaics, similar to inorganic photovoltaics, are generally characterized through current-voltage analysis. This analysis provides multiple device metrics values that are used to understand device performance. One of the most crucial metrics is the Power Conversion Efficiency (PCE).

PCE (η) is proportional to the product of the short-circuit current (J_{SC}), the open circuit voltage (V_{OC}), and the fill factor (FF), all of which can be determined from a current-voltage curve.

$$\eta = \frac{V_{OC} \times J_{SC} \times FF}{P_{in}}$$

Where P_{in} is the incident solar power.

The short circuit current (Jsc), is the maximum photocurrent generation value. It corresponds to the y-intercept value of standard current-voltage curve in which current is plotted along the y-axis and voltage is plotted along the x-axis. Within organic solar cells, the short circuit current can be impacted by a variety of material factors. These include the mobility of charge carriers, the optical absorption profile and general energetic driving forces that lead to a more efficient extraction of charge carriers.

The open circuit voltage (Voc) is the voltage when there is no current running through the device. This corresponds to the x-intercept on a current-voltage curve. Within bulk heterojunction organic photovoltaic devices, this value is highly dependent on HOMO and LUMO energy levels and work functions for the active layer materials.

Since power is the product of voltage and current, the maximum power point occurs when the product between voltage and current is maximized.

The fill factor, FF, can be thought of as the "squareness" of a current voltage curve. It is the quotient of the maximum power value and the product of the open circuit voltage and short circuit current. This is shown in the image above as the ratio of the area of the yellow rectangle to the greater blue rectangle. For organic photovoltaics, this fill factor is essentially a measure of how efficiently generated charges are extracted from the device. This can be thought of as a "competition" between charges transporting through the device, and charges that recombine.

A major issue surrounding polymer solar cells is the low Power Conversion Efficiency (PCE) of fabricated cells. In order to be considered commercially viable, PSCs must be able to achieve at least 10–15% efficiency—this is already much lower than inorganic PVs. However, due to the low cost of polymer solar cells, a 10–15% efficiency is commercially viable.

Recent advances in polymer solar cell performance have resulted from compressing the bandgap to enhance short-circuit current while lowering the Highest Occupied Molecular Orbital (HOMO) to increase open-circuit voltage. However, PSCs still suffer from low fill factors (typically below 70%). However, as of 2013, researchers have been able to fabricate PSCs with fill factors of over 75%. Scientists have been able to accomplish via an inverted BHJ and by using nonconventional donor/acceptor combinations.

Hybrid Solar Cell

Hybrid solar cells combine advantages of both organic and inorganic semiconductors. Hybrid photovoltaics have organic materials that consist of conjugated polymers that absorb light as the donor and transport holes. Inorganic materials in hybrid cells are used as the acceptor and electron transporter in the structure. The hybrid photovoltaic devices have a potential for not only low-cost by roll-to-roll processing but also for scalable solar power conversion.

Solar cells are devices that convert sunlight into electricity by the photovoltaic effect. Electrons in a solar cell absorb photon energy in sunlight which excites them to the conduction band from the valence band. This generates a hole-electron pair, which is separated by a potential barrier (such as a p-n junction), and induces a current. Organic solar cells use organic materials in their active layers. Molecular, polymer, and hybrid organic photovoltaics are the main kinds of organic photovoltaic devices currently studied.

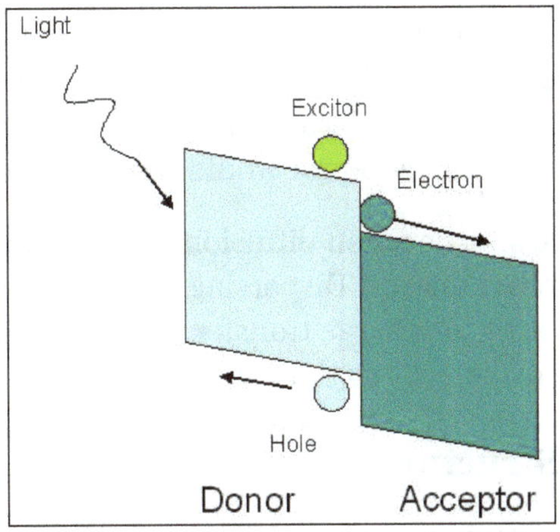

Energy diagram of the donor and acceptor. The conduction band of the acceptor is lower than the LUMO of the polymer, allowing for transfer of the electron.

In hybrid solar cells, an organic material is mixed with a high electron transport material to form the photoactive layer. The two materials are assembled together in a heterojunction-type photoactive layer, which can have a greater power conversion efficiency than a single material. One of the materials acts as the photon absorber and exciton donor. The other material facilitates exciton dissociation at the junction. Charge is transferred and then separated after an exciton created in the donor is delocalized on a donor-acceptor complex.

The acceptor material needs a suitable energy offset to the binding energy of the exciton to the absorber. Charge transfer is favorable if the following condition is satisfied:

$$E_A^A - E_A^D > U_D$$

where superscripts A and D refer to the acceptor and donor respectively, E_A is the electron affinity, and U the coulombic binding energy of the exciton on the donor. An energy diagram of the interface is shown in figure above. In commonly used photovoltaic polymers such as MEH-PPV, the exciton binding energy ranges from 0.3 eV to 1.4 eV.

The energy required to separate the exciton is provided by the energy offset between the LUMOs or conduction bands of the donor and acceptor. After dissociation, the carriers are transported to the respective electrodes through a percolation network.

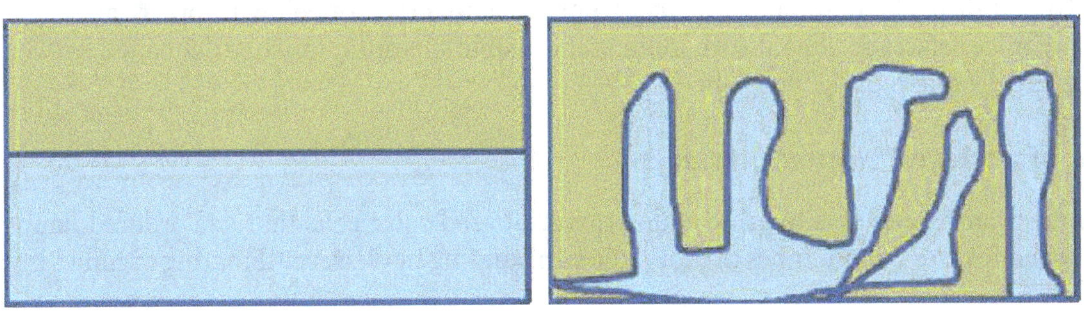

Two different structures of heterojunctions, a) phase separated bi-layer and b) bulk heterojunction. The bulk heterojunction allows for more interfacial contact between the two phases, which is beneficial for the nanoparticle-polymer compound as it provides more surface area for charge transfer.

The average distance an exciton can diffuse through a material before annihilation by recombination is the exciton diffusion length. This is short in polymers, on the order of 5–10 nanometers. The time scale for radiative and non-radiative decay is from 1 picosecond to 1 nanosecond. Excitons generated within this length close to an acceptor would contribute to the photocurrent.

To deal with the problem of the short exciton diffusion length, a bulk heterojunction structure is used rather than a phase-separated bilayer. Dispersing the particles throughout the polymer matrix creates a larger interfacial area for charge transfer to occur. Figure above displays the difference between a bilayer and a bulk heterojunction.

Types of Interfaces and Structures

Controlling the interface of inorganic-organic hybrid solar cells can increase the efficiency of the cells. This increased efficiency can be achieved by increasing the interfacial surface area between the organic and the inorganic to facilitate charge separation and by controlling the nanoscale lengths and periodicity of each structure so that charges are allowed to separate and move toward the appropriate electrode without recombining. The three main nanoscale structures used are mesoporous inorganic films infused with electron-donating organic, alternatining inorganic-organic lamellar structures, and nanowire structures.

Mesoporous Films

Mesoporous films have been used for a relatively high-efficiency hybrid solar cell. The structure of mesoporous thin film solar cells usually includes a porous inorganic that is saturated with organic surfactant. The organic absorbs light, and transfers electrons to the inorganic semiconductor (usually a transparent conducting oxide), which then transfers the electron to the electrode. Problems with these cells include their random ordering and the difficulty of controlling their nanoscale structure to promote charge conduction.

Ordered Lamellar Films

Recently, the use of alternating layers of organic and inorganic compounds has been controlled through electrodeposition-based self-assembly. This is of particular interest because it has been shown that the lamellar structure and periodicity of the alternating organic-inorganic layers can be controlled through solution chemistry. To produce this type of cell with practical efficiencies, larger organic surfactants that absorb more of the visible spectrum must be deposited between the layers of electron-accepting inorganic.

Films of Ordered Nanostructures

Researchers have been able to grow nanostructure-based solar cells that use ordered nanostructures like nanowires or nanotubes of inorganic surrounding by electron-donating organics utilizing self-organization processes. Ordered nanostructures offer the advantage of directed charge transport and controlled phase separation between donor and acceptor materials. The nanowire-based morphology offers reduced internal reflection, facile strain relaxation and increased defect tolerance. The ability to make single-crystalline nanowires on low-cost substrates such as aluminum foil and to relax strain in subsequent layers removes two more major cost hurdles associated with high-efficiency cells. There have been rapid increases in efficiencies of nanowire-based solar cells and they seem to be one of the most promising nanoscale solar hybrid technologies.

Fundamental Challenge Factors

Hybrid cell efficiency must be increased to start large-scale manufacturing. Three factors affect efficiency. First, the bandgap should be reduced to absorb red photons, which contain a significant fraction of the energy in the solar spectrum. Current organic photovoltaics have shown 70% of quantum efficiency for blue photons. Second, contact resistance between each layer in the device should be minimized to offer higher fill factor and power conversion efficiency. Third, charge-carrier mobility should be increased to allow the photovoltaics to have thicker active layers while minimizing carrier recombination and keeping the series resistance of the device low.

Types of Hybrid Solar Cells

Polymer–Nanoparticle Composite

Nanoparticles are a class of semiconductor materials whose size in at least one dimension ranges from 1 to 100 nanometers, on the order of exciton wavelengths. This size control creates quantum confinement and allows for the tuning of optoelectronic properties, such as band gap and electron affinity. Nanoparticles also have a large surface area to volume ratio, which presents more area for charge transfer to occur.

The photoactive layer can be created by mixing nanoparticles into a polymer matrix. Solar devices based on polymer-nanoparticle composites most resemble polymer solar cells. In this case, the nanoparticles take the place of the fullerene based acceptors used in fully organic polymer solar cells. Hybrid solar cells based upon nanoparticles are an area of research interest because nanoparticles have several properties that could make them preferable to fullerenes, such as:

- Fullerenes are synthesized by a combination of a high temperature arc method and continuous gas-phase synthesis, which makes their production difficult and energy intensive. The colloidal synthesis of nanoparticles by contrast is a low temperature process.
- PCBM (a common fullerene acceptor) diffuses during long timespans or when exposed to heat, which can alter the morphology and lower the efficiency of a polymer solar cell. Limited testing of nanoparticle solar cells indicates they may be more stable over time.
- Nanoparticles are more absorbent than fullerenes, meaning more light can be theoretically absorbed in a thinner device.
- Nanoparticle size can affect absorption. This combined with the fact that there are many

possible semiconducting nanoparticles allows for highly customizable bandgaps that can be easily tuned to certain frequencies, which would be advantageous in tandem solar cells.

- Nanoparticles with size near their Bohr radius can generate two excitons when struck by a sufficiently energetic photon.

Structure and Processing

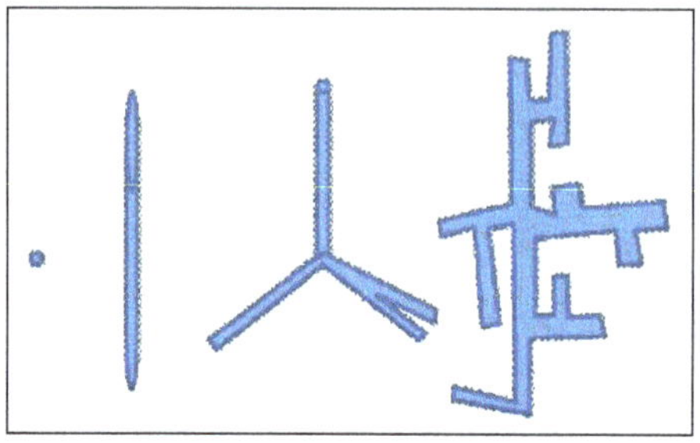

Four different structures of nanoparticles, which have at least 1 dimension in the 1–100 nm range, retaining quantum confinement. Left is a nanocrystal, next to it is nanorod, third is tetrapod, and right is hyperbranched.

For polymers used in this device, hole mobilities are greater than electron mobilities, so the polymer phase is used to transport holes. The nanoparticles transport electrons to the electrode.

The interfacial area between the polymer phase and the nanoparticles needs to be large. This is achieved by dispersing the particles throughout the polymer matrix. However, the nanoparticles need to be interconnected to form percolation networks for electron transport, which occurs by hopping events.

Efficiency is affected by aspect ratio, geometry, and volume fraction of the nanoparticles. Nanoparticle structures include nanocrystals, nanorods, and hyperbranched structures. Figure above contains a picture of each structure. Different structures change the conversion efficiency by effecting nanoparticle dispersion in the polymer and providing pathways for electron transport.

The nanoparticle phase is required to provide a pathway for the electrons to reach the electrode. By using nanorods instead of nanocrystals, the hopping event from one crystal to another can be avoided.

Fabrication methods include mixing the two materials in a solution and spin-coating it onto a substrate, and solvent evaporation (sol-gel). Most of these methods do not involve high-temperature processing. Annealing increases order in the polymer phase, increasing conductivity. However, annealing for too long causes the polymer domain size to increase, eventually making it larger than the exciton diffusion length, and possibly allowing some of the metal from the contact to diffuse into the photoactive layer, reducing the efficiency of the device.

Materials

Inorganic semiconductor nanoparticles used in hybrid cells include CdSe (size ranges from 6–20 nm), ZnO, TiO, and PbS. Common polymers used as photo materials have extensive conjugation and are also hydrophobic. Their efficiency as a photo-material is affected by the HOMO level position and the ionization potential, which directly affects the open circuit voltage and the stability in air. The most common polymers used are P3HT (poly (3-hexylthiophene)), and M3H-PPV (poly[2-methoxy, 5-(2′-ethyl-hexyloxy)-p-phenylenevinylene)]). P3HT has a bandgap of 2.1 eV and M3H-PPV has a bandgap of ~2.4 eV. These values correspond with the bandgap of CdSe, 2.10 eV. The electron affinity of CdSe ranges from 4.4 to 4.7 eV. When the polymer used is MEH-PPV, which has an electron affinity of 3.0 eV, the difference between the electron affinities is large enough to drive electron transfer from the CdSe to the polymer. CdSe also has a high electron mobility (600 $cm^2 \cdot V^{-1} \cdot s^{-1}$).

Performance Values

The highest demonstrated efficiency is 3.2%, based upon a PCPDTBT polymer donor and CdSe nanoparticle acceptor. The device exhibited a short circuit current of 10.1 $mA \cdot cm^{-2}$, an open circuit voltage of .68 V, and a fill factor of .51.

Challenges

Hybrid solar cells need increased efficiencies and stability over time before commercialization is feasible. In comparison to the 2.4% of the CdSe-PPV system, silicon photodevices have power conversion efficiencies greater than 20%.

Problems include controlling the amount of nanoparticle aggregation as the photolayer forms. The particles need to be dispersed in order to maximize interface area, but need to aggregate to form networks for electron transport. The network formation is sensitive to the fabrication conditions. Dead end pathways can impede flow. A possible solution is implementing ordered heterojunctions, where the structure is well controlled.

The structures can undergo morphological changes over time, namely phase separation. Eventually, the polymer domain size will be greater than the carrier diffusion length, which lowers performance.

Even though the nanoparticle bandgap can be tuned, it needs to be matched with the corresponding polymer. The 2.0 eV bandgap of CdSe is larger than an ideal bandgap of 1.4 for absorbance of light.

The nanoparticles involved are typically colloids, which are stabilized in solution by ligands. The ligands decrease device efficiency because they serve as insulators which impede interaction between the donor and nanoparticle acceptor as well as decreasing the electron mobility. Some, but not complete success has been had by exchanging the initial ligands for pyridine or another short chain ligand.

Hybrid solar cells exhibit material properties inferior to those of bulk silicon semiconductors. The carrier mobilities are much smaller than that of silicon. Electron mobility in silicon is 1000 $cm^2 \cdot V^{-1} \cdot s^{-1}$, compared to 600 $cm^2 \cdot V^{-1} \cdot s^{-1}$ in CdSe, and less than 10 $cm^2 \cdot V^{-1} \cdot s^{-1}$ in other quantum dot materials. Hole mobility in MEH-PPV is 0.1 $cm^2 \cdot V^{-1} \cdot s^{-1}$, while in silicon it is 450 $cm^2 \cdot V^{-1} \cdot s^{-1}$.

Carbon Nanotubes

Carbon nanotubes (CNTs) have high electron conductivity, high thermal conductivity, robustness, and flexibility. Field emission displays (FED), strain sensors, and field effect transistors (FET) using CNTs have been demonstrated. Each application shows the potential of CNTs for nanoscale devices and for flexible electronics applications. Photovoltaic applications have also been explored for this material.

Mainly, CNTs have been used as either the photo-induced exciton carrier transport medium impurity within a polymer-based photovoltaic layer or as the photoactive (photon-electron conversion) layer. Metallic CNT is preferred for the former application, while semiconducting CNT is preferred for the later.

Efficient Carrier Transport Medium

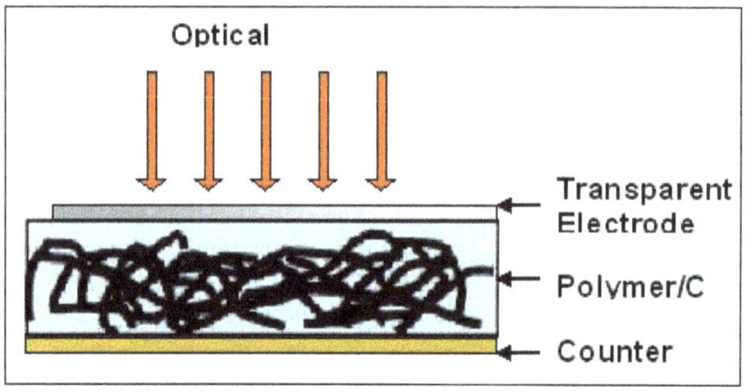

Device diagram for CNT as efficient carrier transport medium.

To increase the photovoltaic efficiency, electron-accepting impurities must be added to the photoactive region. By incorporating CNTs into the polymer, dissociation of the exciton pair can be accomplished by the CNT matrix. The high surface area (~1600 m^2/g) of CNTs offers a good opportunity for exciton dissociation. The separated carriers within the polymer-CNT matrix are transported by the percolation pathways of adjacent CNTs, providing the means for high carrier mobility and efficient charge transfer. The factors of performance of CNT-polymer hybrid photovoltaics are low compared to those of inorganic photovoltaics. SWNT in P3OT semiconductor polymer demonstrated open circuit voltage (V_{oc}) of below 0.94 V, with short circuit current (I_{sc}) of 0.12 mA/cm^2.

Metal nanoparticles may be applied to the exterior of CNTs to increase the exciton separation efficiency. The metal provides a higher electric field at the CNT-polymer interface, accelerating the exciton carriers to transfer them more effectively to the CNT matrix. In this case, V_{oc} = 0.3396 V and I_{sc} = 5.88 mA/cm^2. The fill factor is 0.3876%, and the white light conversion factor 0.775%.

Photoactive Matrix Layer

CNT may be used as a photovoltaic device not only as an add-in material to increase carrier transport, but also as the photoactive layer itself. The semiconducting single walled CNT (SWCNT) is a potentially attractive material for photovoltaic applications for the unique structural and electrical properties. SWCNT has high electric conductivity (100 times that of copper) and shows ballistic carrier transport, greatly decreasing carrier recombination. The bandgap of the SWCNT is inversely proportional to the tube diameter, which means that SWCNT may show multiple direct bandgaps matching the solar spectrum.

A strong built-in electric field in SWCNT for efficient photogenerated electron-hole pair separation has been demonstrated by using two asymmetrical metal electrodes with high and low work functions. The open circuit voltage (V_{oc}) is 0.28 V, with short circuit current (I_{sc}) 1.12 $nA{\cdot}cm^{-2}$ with an incident light source of 8.8 $W \cdot cm^{-2}$. The resulting white light conversion factor is 0.8%.

Challenges

Several challenges must be addressed for CNT to be used in photovoltaic applications. CNT degrades over time in an oxygen-rich environment. The passivation layer required to prevent CNT oxidation may reduce the optical transparency of the electrode region and lower the photovoltaic efficiency.

Challenges as Efficient Carrier Transport Medium

Additional challenges involve the dispersion of CNT within the polymer photoactive layer. The CNT is required to be well dispersed within the polymer matrix to form charge-transfer-efficient pathways between the excitons and the electrode

Challenges as Photoactive Matrix Layer

Challenges of CNT for the photoactive layer include its lack of capability to form a p-n junction, due to the difficulty of doping certain segments of a CNT. (A p-n junction creates an internal built-in potential, providing a pathway for efficient carrier separation within the photovoltaic.) To overcome this difficulty, energy band bending has been done by the use of two electrodes of different work functions. A strong built-in electric field covering the whole SWCNT channel is formed for high-efficiency carrier separation. The oxidation issue with CNT is more critical for this application. Oxidized CNTs have a tendency to become more metallic, and so less useful as a photovoltaic material.

Dye-sensitized

Dye-sensitized solar cells consists of a photo-sensitized anode, an electrolyte, and a photo-electrochemical system. Hybrid solar cells based on dye-sensitized solar cells are formed with inorganic materials (TiO_2) and organic materials.

Materials

Hybrid solar cells based on dye-sensitized solar cells are fabricated by dye-absorbed inorganic materials and organic materials. TiO_2 is the preferred inorganic material since this material is easy to synthesize and acts as a n-type semiconductor due to the donor-like oxygen vacancies. However, titania only absorbs a small fraction of the UV spectrum. Molecular sensitizers (dye molecules) attached to the semiconductor surface are used to collect a greater portion of the spectrum. In the case of titania dye-sensitized solar cells, a photon absorbed by a dye-sensitizer molecule layer induces electron injection into the conduction band of titania, resulting in current flow. However, short diffusion length (diffusivity, $D_n \leq 10^{-4} cm^2/s$) in titania dye-sensitized solar cells decrease the solar-to-energy conversion efficiency. To enhance diffusion length (or carrier lifetime), a variety of organic materials are attached to the titania.

Fabrication Scheme

Dye-sensitized Photoelectrochemical Cell (Grätzel Cell)

TiO_2 nanoparticles are synthesized in several tens of nanometer scales (~100 nm). In order to make a photovoltaic cell, molecular sensitizers (dye molecules) are attached to the titania surface. The dye-absorbed titania is finally enclosed by a liquid electrolyte. This type of dye-sensitized solar cell is also known as a Grätzel cell. Dye-sensitized solar cell has a disadvantage of a short diffusion length. Recently, supermolecular or multifunctional sensitizers have been investigated so as to enhance carrier diffusion length. For example, a dye chromophore has been modified by the addition of secondary electron donors. Minority carriers (holes in this case) diffuse to the attached electron donors to recombine. Therefore, electron-hole recombination is retarded by the physical separation between the dye–cation moiety and the TiO_2 surface, as shown in Figure. Finally, this process raises the carrier diffusion length, resulting in the increase of carrier lifetime.

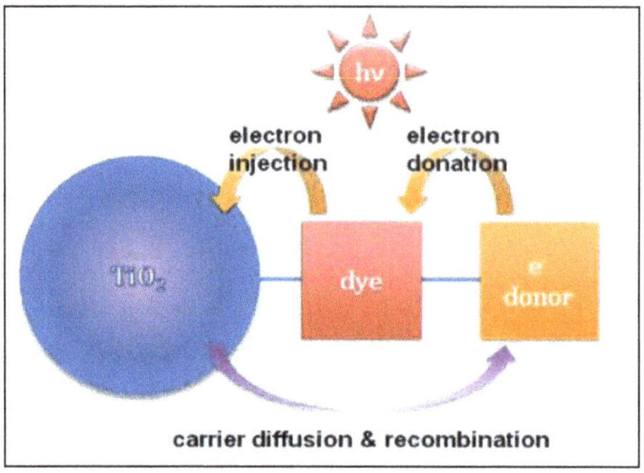

Schematic representation of electron-hole generation and recombination.

Solid-state Dye Sensitized Solar Cell

Mesoporous materials contain pores with diameters between 2 and 50 nm. A dye-sensitized mesoporous film of TiO_2 can be used for making photovoltaic cells and this solar cell is called a 'solid-state dye sensitized solar cell'. The pores in mesoporous TiO_2 thin film are filled with a solid hole-conducting material such as p-type semiconductors or organic hole conducting material. Replacing the liquid electrolyte in Grätzel's cells with a solid charge-transport material can be beneficial. The process of electron-hole generation and recombination is the same as Grätzel cells. Electrons are injected from photoexcited dye into the conduction band of titania and holes are transported by a solid charge transport electrolyte to an electrode. Many organic materials have been tested to obtain a high solar-to-energy conversion efficiency in dye synthesized solar cells based on mesoporous titania thin film.

Efficiency Factors

Efficiency factors demonstrated for dye-sensitized solar cells are:

Parameters	Types of dye sensitized solar cells	
	Grätzel cell	Solid-state
Efficiency (%)	~ 10–11	~ 4
Voc (V)	~ 0.7	~ 0.40
Jsc (mA/cm2)	~ 20	~ 9.10
Fill factor	~ 0.67	~ 0.6

Challenges

Liquid organic electrolytes contain highly corrosive iodine, leading to problems of leakage, sealing, handling, dye desorption, and maintenance. Much attention is now focused on the electrolyte to address these problems.

For solid-state dye sensitized solar cells, the first challenge originates from disordered titania mesoporous structures. Mesoporous titania structures should be fabricated with well-ordered titania structures of uniform size (~ 10 nm). The second challenge comes from developing the solid electrolyte, which is required to have these properties:

- The electrolyte should be transparent to the visible spectrum (wide band gap).
- Fabrication should be possible for depositing the solid electrolyte without degrading the dye molecule layer on titania.
- The LUMO of the dye molecule should be higher than the conduction band of titania.
- Several p-type semiconductors tend to crystallize inside the mesoporous titania films, destroying the dye molecule-titania contact. Therefore, the solid electrolyte needs to be stable during operation.

Nanostructured Inorganic — Small Molecules

In 2008, scientists were able to create a nanostructured lamellar structure that provides an ideal design for bulk heterojunction solar cells. The observed structure is composed of ZnO and small, conducting organic molecules, which co-assemble into alternating layers of organic and inorganic components. This highly organized structure, which is stabilized by π-π stacking between the organic molecules, allows for conducting pathways in both the organic and inorganic layers. The thicknesses of the layers (about 1–3 nm) are well within the exciton diffusion length, which ideally minimizes recombination among charge carriers. This structure also maximizes the interface between the inorganic ZnO and the organic molecules, which enables a high chromophore loading density within the structure. Due to the choice of materials, this system is non-toxic and environmentally friendly, unlike many other systems which use lead or cadmium.

Although this system has not yet been incorporated into a photovoltaic device, preliminary photoconductivity measurements have shown that this system exhibits among the highest values measured for organic, hybrid, and amorphous silicon photoconductors, and so, offers promise in creating efficient hybrid photovoltaic devices.

Quantum Dot Solar Cell

A quantum dot solar cell (QDSC) is a solar cell design that uses quantum dots as the absorbing photovoltaic material. It attempts to replace bulk materials such as silicon, copper indium gallium selenide (CIGS) or cadmium telluride (CdTe). Quantum dots have bandgaps that are tunable across a wide range of energy levels by changing their size. In bulk materials, the bandgap is fixed by the choice of material(s). This property makes quantum dots attractive for multi-junction solar cells, where a variety of materials are used to improve efficiency by harvesting multiple portions of the solar spectrum. As of 2019, efficiency exceeds 16.5%.

Spin-cast quantum dot solar cell built by the Sargent Group at the University of Toronto. The metal disks on the front surface are the electrical connections to the layers below.

Solar Cell Concepts

In a conventional solar cell, light is absorbed by a semiconductor, producing an electron-hole (e-h) pair; the pair may be bound and is referred to as an exciton. This pair is separated by an internal electrochemical potential (present in p-n junctions or Schottky diodes) and the resulting flow of electrons and holes creates electric current. The internal electrochemical potential is created by doping one part of semiconductor interface with atoms that act as electron donors (n-type doping) and another with electron acceptors (p-type doping) that results in a p-n junction. Generation of an e-h pair requires that the photons have energy exceeding the bandgap of the material. Effectively, photons with energies lower than the bandgap do not get absorbed, while those that are higher can quickly (within about 10^{-13} s) thermalize to the band edges, reducing output. The former limitation reduces current, while the thermalization reduces the voltage. As a result, semiconductor cells suffer a trade-off between voltage and current (which can be in part alleviated by using multiple junction implementations). The detailed balance calculation shows that this efficiency can not exceed 33% if one uses a single material with an ideal bandgap of 1.34 eV for a solar cell.

The band gap (1.34 eV) of an ideal single-junction cell is close to that of silicon (1.1 eV), one of the many reasons that silicon dominates the market. However, silicon's efficiency is limited to about 30% (Shockley–Queisser limit). It is possible to improve on a single-junction cell by vertically stacking cells with different bandgaps – termed a "tandem" or "multi-junction" approach. The same analysis shows that a two layer cell should have one layer tuned to 1.64 eV and the other to 0.94 eV, providing a theoretical performance of 44%. A three-layer cell should be tuned to 1.83, 1.16 and 0.71 eV, with an efficiency of 48%. An "infinity-layer" cell would have a theoretical efficiency of 86%, with other thermodynamic loss mechanisms accounting for the rest.

Traditional (crystalline) silicon preparation methods do not lend themselves to this approach due to lack of bandgap tunability. Thin-films of amorphous silicon, which due to a relaxed requirement in crystal momentum preservation can achieve direct bandgaps and intermixing of carbon, can tune the bandgap, but other issues have prevented these from matching the performance of traditional cells. Most tandem-cell structures are based on higher performance semiconductors, notably indium gallium arsenide (InGaAs). Three-layer InGaAs/GaAs/InGaP cells (bandgaps 0.94/1.42/1.89 eV) hold the efficiency record of 42.3% for experimental examples.

However, the QDSCs suffer from weak absorption and the contribution of the light absorption at room temperature is marginal. This can be addressed by utilizing multibranched Au nanostars.

Quantum dots

Quantum dots are semiconducting particles that have been reduced below the size of the Exciton Bohr radius and due to quantum mechanics considerations, the electron energies that can exist within them become finite, much alike energies in an atom. Quantum dots have been referred to as "artificial atoms". These energy levels are tuneable by changing their size, which in turn defines the bandgap. The dots can be grown over a range of sizes, allowing them to express a variety of bandgaps without changing the underlying material or construction techniques. In typical wet chemistry preparations, the tuning is accomplished by varying the synthesis duration or temperature.

The ability to tune the bandgap makes quantum dots desirable for solar cells. For the sun's photon distribution spectrum, the Shockley-Queisser limit indicates that the maximum solar conversion efficiency occurs in a material with a band gap of 1.34 eV. However, materials with lower band gaps will be better suited to generate electricity from lower-energy photons (and vice versa). Single junction implementations using lead sulfide (PbS) colloidal quantum dots (CQD) have bandgaps that can be tuned into the far infrared, frequencies that are typically difficult to achieve with traditional solar cells. Half of the solar energy reaching the Earth is in the infrared, most in the near infrared region. A quantum dot solar cell makes infrared energy as accessible as any other.

Moreover, CQD offer easy synthesis and preparation. While suspended in a colloidal liquid form they can be easily handled throughout production, with a fumehood as the most complex equipment needed. CQD are typically synthesized in small batches, but can be mass-produced. The dots can be distributed on a substrate by spin coating, either by hand or in an automated process. Large-scale production could use spray-on or roll-printing systems, dramatically reducing module construction costs.

Production

Early examples used costly molecular beam epitaxy processes. However, the lattice mismatch results in accumulation of strain and thus generation of defects, restricting the number of stacked layers. Droplet epitaxy growth technique shows its advantages on the fabrication of strain-free QDs. Alternatively, less expensive fabrication methods were later developed. These use wet chemistry (for CQD) and subsequent solution processing. Concentrated nanoparticle solutions are stabilized by long hydrocarbon ligands that keep the nanocrystals suspended in solution.

To create a solid, these solutions are cast down and the long stabilizing ligands are replaced with short-chain crosslinkers. Chemically engineering the nanocrystal surface can better passivate the nanocrystals and reduce detrimental trap states that would curtail device performance by means of carrier recombination. This approach produces an efficiency of 7.0%.

A more recent study uses different ligands for different functions by tuning their relative band alignment to improve the performance to 8.6%. The cells were solution-processed in air at room-temperature and exhibited air-stability for more than 150 days without encapsulation.

In 2014 the use of iodide as a ligand that does not bond to oxygen was introduced. This maintains stable n- and p-type layers, boosting the absorption efficiency, which produced power conversion efficiency up to 8%.

Plasmonic Solar Cell

A plasmonic-enhanced solar cell, commonly referred to simply as plasmonic solar cell, is a type of solar cell (including thin-film, crystalline silicon, amorphous silicon, and other types of cells) that converts light into electricity with the assistance of plasmons, but where the photovoltaic effect occurs in another material.

A direct plasmonic solar cell is a solar cell that converts light into electricity using plasmons as the active, photovoltaic material.

The thickness varies from that of traditional silicon PV, to less than 2 μm thick and theoretically could be as thin as 100 nm. They can use substrates which are cheaper than silicon, such as glass, plastic or steel. One of the challenges for thin film solar cells is that they do not absorb as much light as thicker solar cells made with materials with the same absorption coefficient. Methods for light trapping are important for thin film solar cells. Plasmonic-enhanced cells improve absorption by scattering light using metal nano-particles excited at their surface plasmon resonance. Interestingly, plasmonic core-shell nanoparticles located in the front of the thin film solar cells can aid weak absorption of Si solar cells in the near-infrared region—the fraction of light scattered into the substrate and the maximum optical path length enhancement can be as high as 0.999 and 3133. Incoming light at the plasmon resonance frequency induces electron oscillations at the surface of the nanoparticles. The oscillation electrons can then be captured by a conductive layer producing an electrical current. The voltage produced is dependent on the bandgap of the conductive layer and the potential of the electrolyte in contact with the nanoparticles. There is still considerable research necessary to enable the technology to reach its full potential and commercialization of plasmonic-enhanced solar cells.

Design

The design for plasmonic-enhanced solar cells varies depending on the method being used to trap and scatter light across the surface and through the material.

Nanoparticle Cells

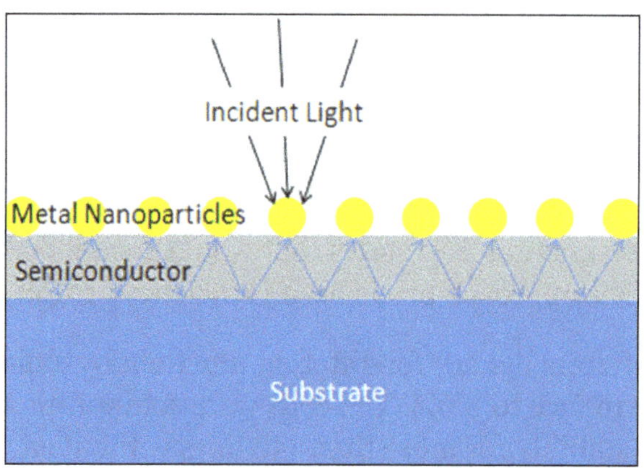

PSC using metal nano-particles.

A common design is to deposit metal nano-particles on the top surface of the surface of the solar cell. When light hits these metal nano-particles at their surface plasmon resonance, the light is scattered in many different directions. This allows light to travel along the solar cell and bounce between the substrate and the nano-particles enabling the solar cell to absorb more light. The concentrated near field intensity induced by localized surface plasmon of the metal nanoparticles will promote the optical absorption of semiconductors. Recently, the plasmonic asymmetric modes of nano particles have found to favor the broadband optical absorption and promote the electrical properties of solar cells. The simultaneously plasmon-optical and plasmon-electrical effects of nanoparticles reveal a promising feature of nanoparticle plasmon.

Recently, the core (metal)-shell (dielectric) nanoparticle has demonstrated a zero backward scattering with enhanced forward scattering on Si substrate when surface plasmon is located in front of a solar cell. The core-shell nanoparticles can support simultaneously both electric and magnetic resonances, demonstrating entirely new properties when compared with bare metallic nanoparticles if the resonances are properly engineered.

Metal Film Cells

Other methods utilizing surface plasmons for harvesting solar energy are available. One other type of structure is to have a thin film of silicon and a thin layer of metal deposited on the lower surface. The light will travel through the silicon and generate surface plasmons on the interface of the silicon and metal. This generates electric fields inside of the silicon since electric fields do not travel very far into metals. If the electric field is strong enough, electrons can be moved and collected to produce a photocurrent. The thin film of metal in this design must have nanometer sized grooves which act as waveguides for the incoming light in order to excite as many photons in the silicon thin film as possible.

Principles

General

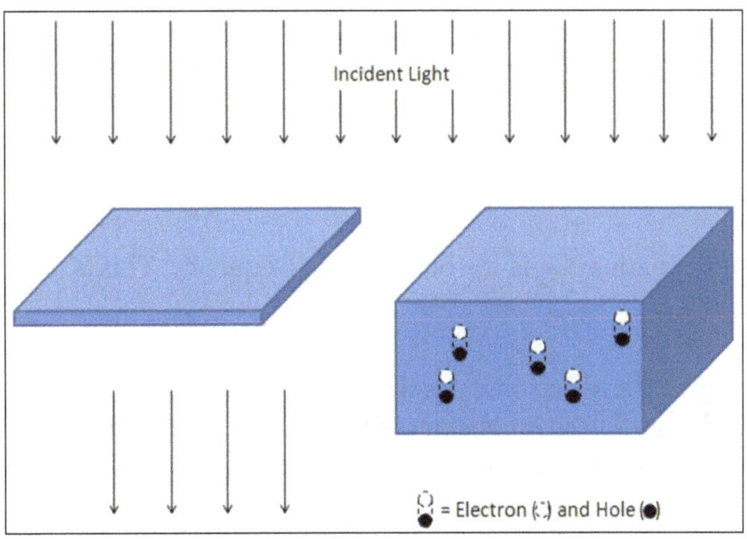

Thin film SC (left) and Typical SC (right).

When a photon is excited in the substrate of a solar cell, an electron and hole are separated. Once the electrons and holes are separated, they will want to recombine since they are of opposite charge. If the electrons can be collected prior to this happening they can be used as a current for an external circuit. Designing the thickness of a solar cell is always a trade-off between minimizing this recombination (thinner layers) and absorbing more photons (thicker layer).

Nano-particles

Scattering and Absorption

The basic principles for the functioning of plasmonic-enhanced solar cells include scattering and absorption of light due to the deposition of metal nano-particles. Silicon does not absorb light very well. For this reason, more light needs to be scattered across the surface in order to increase the absorption. It has been found that metal nano-particles help to scatter the incoming light across the surface of the silicon substrate. The equations that govern the scattering and absorption of light can be shown as:

- $C_{scat} = \frac{1}{6\pi}(\frac{2\pi}{\lambda})^4 |\alpha|^2$

  This shows the scattering of light for particles which have diameters below the wavelength of light.

- $C_{abs} = \frac{2\pi}{\lambda}\text{Im}[\alpha]$

  This shows the absorption for a point dipole model.

- $\alpha = 3V\left[\frac{\epsilon_p/\epsilon_m - 1}{\epsilon_p/\epsilon_m + 2}\right]$

  This is the polarizability of the particle. V is the particle volume. ϵ_p is the dielectric function of the particle. ϵ_m is the dielectric function of the embedding medium. When $\epsilon_p = -2\epsilon_m$ the polarizability of the particle becomes large. This polarizability value is known as the surface plasmon resonance. The dielectric function for metals with low absorption can be defined as:

- $\epsilon = 1 - \frac{\omega_p^2}{\omega^2 + i\gamma\omega}$

  In the previous equation, ω_p is the bulk plasma frequency. This is defined as:

- $\omega_p^2 = Ne^2 / m\epsilon_0$

  N is the density of free electrons, e is the electronic charge and m is the effective mass of an electron. ϵ_0 is the dielectric constant of free space. The equation for the surface plasmon resonance in free space can therefore be represented by:

- $\alpha = 3V\frac{\omega_p^2}{\omega_p^2 - 3\omega^2 - i\gamma\omega}$

Many of the plasmonic solar cells use nano-particles to enhance the scattering of light. These nano-particles take the shape of spheres, and therefore the surface plasmon resonance frequency for spheres is desirable. By solving the previous equations, the surface plasmon resonance frequency for a sphere in free space can be shown as:

- $\omega_{sp} = \sqrt{3}\omega_p$

As an example, at the surface plasmon resonance for a silver nanoparticle, the scattering cross-section is about 10x the cross-section of the nanoparticle. The goal of the nano-particles is to trap light on the surface of the SC. The absorption of light is not important for the nanoparticle, rather, it is important for the SC. One would think that if the nanoparticle is increased in size, then the scattering cross-section becomes larger. This is true, however, when compared with the size of the nanoparticle, the ratio ($CS_{scat} / CS_{particle}$) is reduced. Particles with a large scattering cross section tend to have a broader plasmon resonance range.

Wavelength Dependence

Surface plasmon resonance mainly depends on the density of free electrons in the particle. The order of densities of electrons for different metals is shown below along with the type of light which corresponds to the resonance.

- Aluminum: Ultra-violet
- Silver: Ultra-violet
- Gold: Visible
- Copper: Visible

If the dielectric constant for the embedding medium is varied, the resonant frequency can be shifted. Higher indexes of refraction will lead to a longer wavelength frequency.

Light Trapping

The metal nano-particles are deposited at a distance from the substrate in order to trap the light between the substrate and the particles. The particles are embedded in a material on top of the substrate. The material is typically a dielectric, such as silicon or silicon nitride. When performing experiment and simulations on the amount of light scattered into the substrate due to the distance between the particle and substrate, air is used as the embedding material as a reference. It has been found that the amount of light radiated into the substrate decreases with distance from the substrate. This means that nano-particles on the surface are desirable for radiating light into the substrate, but if there is no distance between the particle and substrate, then the light is not trapped and more light escapes.

The surface plasmons are the excitations of the conduction electrons at the interface of metal and the dielectric. Metallic nano-particles can be used to couple and trap freely propagating plane waves into the semiconductor thin film layer. Light can be folded into the absorbing layer to increase the absorption. The localized surface plasmons in metal nano-particles and the surface plasmon polaritons at the interface of metal and semiconductor are of interest in the current research. In recent reported papers, the shape and size of the metal nano-particles are key factors to determine the incoupling efficiency. The smaller particles have larger incoupling efficiency due to the enhanced near-field coupling. However, very small particles suffer from large ohmic losses.

Recently, the plasmonic asymmetric modes of nano particles have found to favor the broadband optical absorption and promote the electrical properties of solar cells. The simultaneously plasmon-optical and plasmon-electrical effects of nanoparticles reveal a promising feature of nanoparticle plasmon.

Metal Film

As light is incident upon the surface of the metal film, it excites surface plasmons. The surface plasmon frequency is specific for the material, but through the use of gratings on the surface of the film, different frequencies can be obtained. The surface plasmons are also preserved through the use of waveguides as they make the surface plasmons easier to travel on the surface and the losses due to resistance and radiation are minimized. The electric field generated by the surface plasmons influences the electrons to travel toward the collecting substrate.

Materials

First Generation	Second Generation	Third Generation
Single-crystal silicon	CuInSe2	Gallium Indium Phosphide
Multicrystalline silicon	Amorphous silicon	Gallium Indium Arsenide
Polycrystalline silicon	Thin film crystalline Si	Germanium

Applications

The applications for plasmonic-enhanced solar cells are endless. The need for cheaper and more efficient solar cells is huge. In order for solar cells to be considered cost effective, they need to provide energy for a smaller price than that of traditional power sources such as coal and gasoline. The movement toward a more green world has helped to spark research in the area of plasmonic-enhanced solar cells. Currently, solar cells cannot exceed efficiencies of about 30% (First Generation). With new technologies (Third Generation), efficiencies of up to 40-60% can be expected. With a reduction of materials through the use of thin film technology (Second Generation), prices can be driven lower.

Certain applications for plasmonic-enhanced solar cells would be for space exploration vehicles. A main contribution for this would be the reduced weight of the solar cells. An external fuel source would also not be needed if enough power could be generated from the solar cells. This would drastically help to reduce the weight as well.

Solar cells have a great potential to help rural electrification. An estimated two million villages near the equator have limited access to electricity and fossil fuels and that approximately 25% of people in the world do not have access to electricity. When the cost of extending power grids, running rural electricity and using diesel generators is compared with the cost of solar cells, many times the solar cells win. If the efficiency and cost of the current solar cell technology is decreased even further, then many rural communities and villages around the world could obtain electricity when current methods are out of the question. Specific applications for rural communities would be water pumping systems, residential electric supply and street lights. A particularly interesting application would be for health systems in countries where motorized vehicles are not overly abundant. Solar cells could be used to provide the power to refrigerate medications in coolers during transport.

Solar cells could also provide power to lighthouses, buoys, or even battleships out in the ocean. Industrial companies could use them to power telecommunications systems or monitoring and control systems along pipelines or other system.

If the solar cells could be produced on a large scale and be cost effective then entire power stations could be built in order to provide power to the electrical grids. With a reduction in size, they could be implemented on both commercial and residential buildings with a much smaller footprint. They might not even seem like an eyesore.

Other areas are in hybrid systems. The solar cells could help to power high consumption devices such as automobiles in order to reduce the amount of fossil fuels used and to help improve the environmental conditions of the earth.

In consumer electronics devices, solar cells could be used to replace batteries for low power electronics. This would save everyone a lot of money and it would also help to reduce the amount of waste going into landfills.

Advancements

Choice of Plasmonic Metal Nano-particles

Proper choice of plasmatic metal nano-particles is crucial for the maximum light absorption in the active layer. Front surface located nano-particles Ag and Au are the most widely used materials due to their surface plasmon resonances located in the visible range and therefore interact more strongly with the peak solar intensity. However, such noble metal nano-particles always introduce reduced light coupling into Si at the short wavelengths below the surface plasmon resonance due to the detrimental Fano effect, i.e. the destructive interference between the scattered and unscattered light. Moreover, the noble metal nano-particles are impractical to implement for large-scale solar cell manufacture due to their high cost and scarcity in the earth's crust. Recently, Zhang et al. have demonstrated the low cost and earth abundant materials Al nano-particles to be able to outperform the widely used Ag and Au nano-particles. Al nano-particles, with their surface plasmon resonances located in the UV region below the desired solar spectrum edge at 300 nm, can avoid the reduction and introduce extra enhancement in the shorter wavelength range.

Shape Choice of Nano-particles

Shape
Nanosphere
Nanostar
Core-shell nanoparticle
Nanodisk
Nanocavity
Nanovoid
Nucleated nanoparticle
Nanocage
Core-shell nanoparticle

Light Trapping

Being able to concentrate and scatter light across the surface of the plasmonic-enhanced solar cell will help to increase efficiencies. Recently, research at Sandia National Laboratories has discovered a photonic waveguide which collects light at a certain wavelength and traps it within the structure. This new structure can contain 95% of the light that enters it compared to 30% for other traditional waveguides. It can also direct the light within one wavelength which is ten times greater than traditional waveguides. The wavelength this device captures can be selected by changing the structure of the lattice which comprises the structure. If this structure is used to trap light and keep it in the structure until the solar cell can absorb it, the efficiency of the solar cell could be increased dramatically.

Absorption

Another recent advancement in plasmonic-enhanced solar cells is using other methods to aid in the absorption of light. One way being researched is the use of metal wires on top of the substrate to scatter the light. This would help by utilizing a larger area of the surface of the solar cell for light scattering and absorption. The danger in using lines instead of dots would be creating a reflective layer which would reject light from the system. This is very undesirable for solar cells. This would be very similar to the thin metal film approach, but it also utilizes the scattering effect of the nano-particles. Yue, et al. used a type of new materials, called topological insulators, to increase the absorption of ultrathin a-Si solar cells. The topological insulator nanostructure has intrinsically core-shell configuration. The core is dielectric and has ultrahigh refractive index. The shell is metallic and support surface plasmon resonances. Through integrating the nanocone arrays into a-Si thin film solar cells, up to 15% enhancement of light absorption was predicted in the ultraviolet and visible ranges.

Third Generation

The goal of third generation solar cells is to increase the efficiency using second generation solar cells (thin film) and using materials that are found abundantly on earth. This has also been a goal of the thin film solar cells. With the use of common and safe materials, third generation solar cells should be able to be manufactured in mass quantities further reducing the costs. The initial costs would be high in order to produce the manufacturing processes, but after that they should be cheap. The way third generation solar cells will be able to improve efficiency is to absorb a wider range of frequencies. The current thin film technology has been limited to one frequency due to the use of single band gap devices.

Multiple Energy Levels

The idea for multiple energy level solar cells is to basically stack thin film solar cells on top of each other. Each thin film solar cell would have a different band gap which means that if part of the solar spectrum was not absorbed by the first cell then the one just below would be able to absorb part of the spectrum. These can be stacked and an optimal band gap can be used for each cell in order to produce the maximum amount of power. Options for how each cell is connected are available, such as serial or parallel. The serial connection is desired because the output of the solar cell would just be two leads.

The lattice structure in each of the thin film cells needs to be the same. If it is not then there will be losses. The processes used for depositing the layers are complex. They include Molecular Beam Epitaxy and Metal Organic Vapour Phase Epitaxy. The current efficiency record is made with this process but doesn't have exact matching lattice constants. The losses due to this are not as effective because the differences in lattices allows for more optimal band gap material for the first two cells. This type of cell is expected to be able to be 50% efficient.

Lower quality materials that use cheaper deposition processes are being researched as well. These devices are not as efficient, but the price, size and power combined allow them to be just as cost effective. Since the processes are simpler and the materials are more readily available, the mass production of these devices is more economical.

Hot Carrier Cells

A problem with solar cells is that the high energy photons that hit the surface are converted to heat. This is a loss for the cell because the incoming photons are not converted into usable energy. The idea behind the hot carrier cell is to utilize some of that incoming energy which is converted to heat. If the electrons and holes can be collected while hot, a higher voltage can be obtained from the cell. The problem with doing this is that the contacts which collect the electrons and holes will cool the material. Thus far, keeping the contacts from cooling the cell has been theoretical. Another way of improving the efficiency of the solar cell using the heat generated is to have a cell which allows lower energy photons to excite electron and hole pairs. This requires a small bandgap. Using a selective contact, the lower energy electrons and holes can be collected while allowing the higher energy ones to continue moving through the cell. The selective contacts are made using a double barrier resonant tunneling structure. The carriers are cooled which they scatter with phonons. If a material with a large bandgap of phonons then the carriers will carry more of the heat to the contact and it won't be lost in the lattice structure. One material which has a large bandgap of phonons is indium nitride. The hot carrier cells are in their infancy but are beginning to move toward the experimental stage.

Plasmonic-electrical Solar Cells

Having unique features of tunable resonances and unprecedented near-field enhancement, plasmon is an enabling technique for light management. Recently, performances of thin-film solar cells have been pronouncedly improved by introducing metallic nanostructures. The improvements are mainly attributed to the plasmonic-optical effects for manipulating light propagation, absorption, and scattering. The plasmonic-optical effects could: (1) boost optical absorption of active materials; (2) spatially redistribute light absorption at the active layer due to the localized near-field enhancement around metallic nanostructures. Except for the plasmonic-optical effects, the effects of plasmonically modified recombination, transport and collection of photocarriers (electrons and holes), hereafter named plasmonic-electrical effects, have been proposed by Sha, etal. For boosting device performance, they conceived a general design rule, tailored to arbitrary electron to hole mobility ratio, to decide the transport paths of photocarriers. The design rule suggests that electron to hole transport length ratio should be balanced with electron to hole mobility ratio. In other words, the transport time of electrons and holes (from initial generation sites to corresponding electrodes) should be the same. The general design rule can be realized by spatially redistributing light absorption at the active layer of devices (with the plasmonic-electrical effect). They also demonstrated the breaking of space charge limit in plasmonic-electrical organic solar cell. Recently, the plasmonic asymmetric modes of nano particles have found to favor the broadband optical absorption and promote the electrical properties of solar cells. The simultaneously plasmon-optical and plasmon-electrical effects of nanoparticles reveal a promising feature of nanoparticle plasmon.

Ultra-thin Plasmonic Wafer Solar Cells

Reducing the silicon wafer thickness at a minimized efficiency loss represents a mainstream trend in increasing the cost-effectiveness of wafer-based solar cells. Recently, Zhang et al. have demonstrated that, using the advanced light trapping strategy with a properly designed nano-particle architecture, the wafer thickness can be dramatically reduced to only around 1/10 of the current thickness (180 μm) without any solar cell efficiency loss at 18.2%. Nano-particle integrated ultra-thin solar cells with only 3% of the current wafer thickness can potentially achieve 15.3% efficiency combining the absorption enhancement with the benefit of thinner wafer induced open circuit voltage increase. This represents a 97% material saving with only 15% relative efficiency loss. These results demonstrate the feasibility and prospect of achieving high-efficiency ultra-thin silicon wafer cells with plasmonic light trapping.

Direct Plasmonic Solar Cells

The development of direct plasmonic solar cells that use plasmonic nanoparticles directly as light absorbers is much more recent than plasmonic-enhanced cells.

In 2013 it was confirmed that hot carriers in plasmonic nanoparticles can be generated by excitation of localized surface plasmon resonance. The hot electrons were shown to be injected into a TiO_2 conduction band, confirming their usability for light conversion to electricity. In 2019 another article was published describing how the hot electrons counterpart, the hot holes, can also be injected into a p-type semiconductor. This separation of charges enables direct use of plasmonic nanoparticles as light absorbers in photovoltaic cells.

A spin-off company from Uppsala university, Peafowl Solar Power, is developing the direct plasmonic solar cell technology for commercial applications such as transparent solar cells for dynamic glass.

Dye-sensitized Solar Cell

A selection of dye-sensitized solar cells.

A dye-sensitized solar cell (DSSC, DSC, DYSC or Grätzel cell) is a low-cost solar cell belonging to the group of thin film solar cells. It is based on a semiconductor formed between a photo-sensitized anode and an electrolyte, a *photoelectrochemical* system. The modern version of a dye solar cell, also known as the Grätzel cell, was originally co-invented in 1988 by Brian O'Regan and Michael Grätzel at UC Berkeley and this work was later developed by the aforementioned scientists at the École Polytechnique Fédérale de Lausanne until the publication of the first high efficiency DSSC in 1991. Michael Grätzel has been awarded the 2010 Millennium Technology Prize for this invention.

The DSSC has a number of attractive features; it is simple to make using conventional roll-printing techniques, is semi-flexible and semi-transparent which offers a variety of uses not applicable to glass-based systems, and most of the materials used are low-cost. In practice it has proven difficult to eliminate a number of expensive materials, notably platinum and ruthenium, and the liquid electrolyte presents a serious challenge to making a cell suitable for use in all weather. Although its conversion efficiency is less than the best thin-film cells, in theory its price/performance ratio should be good enough to allow them to compete with fossil fuel electrical generation by achieving grid parity. Commercial applications, which were held up due to chemical stability problems, are forecast in the European Union Photovoltaic Roadmap to significantly contribute to renewable electricity generation by 2020.

Current Technology: Semiconductor Solar Cells

In a traditional solid-state semiconductor, a solar cell is made from two doped crystals, one doped with n-type impurities (n-type semiconductor), which add additional free conduction band electrons, and the other doped with p-type impurities (p-type semiconductor), which add additional electron holes. When placed in contact, some of the electrons in the n-type portion flow into the p-type to "fill in" the missing electrons, also known as electron holes. Eventually enough electrons will flow across the boundary to equalize the Fermi levels of the two materials. The result is a region at the interface, the p-n junction, where charge carriers are depleted and/or accumulated on each side of the interface. In silicon, this transfer of electrons produces a potential barrier of about 0.6 to 0.7 V.

When placed in the sun, photons of the sunlight can excite electrons on the p-type side of the semiconductor, a process known as photoexcitation. In silicon, sunlight can provide enough energy to push an electron out of the lower-energy valence band into the higher-energy conduction band. As the name implies, electrons in the conduction band are free to move about the silicon. When a load is placed across the cell as a whole, these electrons will flow out of the p-type side into the n-type side, lose energy while moving through the external circuit, and then flow back into the p-type material where they can once again re-combine with the valence-band hole they left behind. In this way, sunlight creates an electric current.

In any semiconductor, the band gap means that only photons with that amount of energy, or more, will contribute to producing a current. In the case of silicon, the majority of visible light from red to violet has sufficient energy to make this happen. Unfortunately higher energy photons, those at the blue and violet end of the spectrum, have more than enough energy to cross the band gap; although some of this extra energy is transferred into the electrons, the majority of it is wasted as heat. Another issue is that in order to have a reasonable chance of capturing a photon, the n-type layer has to be fairly thick. This also increases the chance that a freshly ejected electron will meet up with a previously created hole in the material before reaching the p-n junction. These effects produce an upper limit on the efficiency of silicon solar cells, currently around 12 to 15% for common modules and up to 25% for the best laboratory cells (33.16% is the theoretical maximum efficiency for single band gap solar cells).

By far the biggest problem with the conventional approach is cost; solar cells require a relatively thick layer of doped silicon in order to have reasonable photon capture rates, and silicon processing is expensive. There have been a number of different approaches to reduce this cost over the last decade, notably the thin-film approaches, but to date they have seen limited application due to a variety of practical problems. Another line of research has been to dramatically improve efficiency through the multi-junction approach, although these cells are very high cost and suitable only for large commercial deployments. In general terms the types of cells suitable for rooftop deployment have not changed significantly in efficiency, although costs have dropped somewhat due to increased supply.

Dye-sensitized Solar Cells

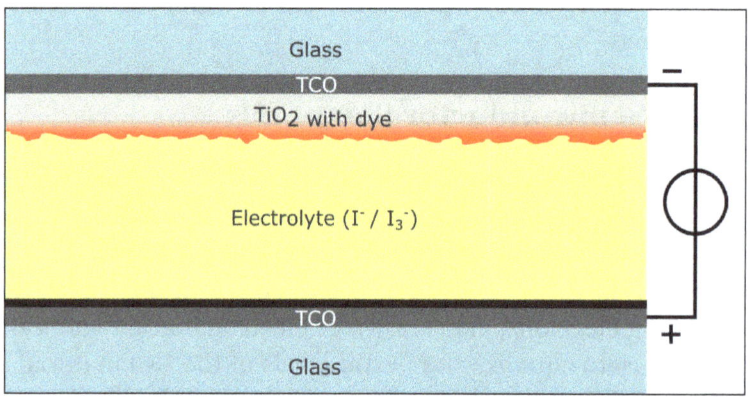

Type of cell made at the EPFL by Grätzel and O'Regan.

In the late 1960s it was discovered that illuminated organic dyes can generate electricity at oxide electrodes in electrochemical cells. In an effort to understand and simulate the primary processes in photosynthesis the phenomenon was studied at the University of California at Berkeley with chlorophyll extracted from spinach (bio-mimetic or bionic approach). On the basis of such experiments electric power generation via the dye sensitization solar cell (DSSC) principle was demonstrated and discussed in 1972. The instability of the dye solar cell was identified as a main challenge. Its efficiency could, during the following two decades, be improved by optimizing the porosity of the electrode prepared from fine oxide powder, but the instability remained a problem.

A modern DSSC is composed of a porous layer of titanium dioxide nanoparticles, covered with a molecular dye that absorbs sunlight, like the chlorophyll in green leaves. The titanium dioxide is immersed under an electrolyte solution, above which is a platinum-based catalyst. As in a conventional alkaline battery, an anode (the titanium dioxide) and a cathode (the platinum) are placed on either side of a liquid conductor (the electrolyte).

Sunlight passes through the transparent electrode into the dye layer where it can excite electrons that then flow into the titanium dioxide. The electrons flow toward the transparent electrode where they are collected for powering a load. After flowing through the external circuit, they are re-introduced into the cell on a metal electrode on the back, flowing into the electrolyte. The electrolyte then transports the electrons back to the dye molecules.

Dye-sensitized solar cells separate the two functions provided by silicon in a traditional cell design. Normally the silicon acts as both the source of photoelectrons, as well as providing the electric field to separate the charges and create a current. In the dye-sensitized solar cell, the bulk of the semiconductor is used solely for charge transport, the photoelectrons are provided from a separate photosensitive dye. Charge separation occurs at the surfaces between the dye, semiconductor and electrolyte.

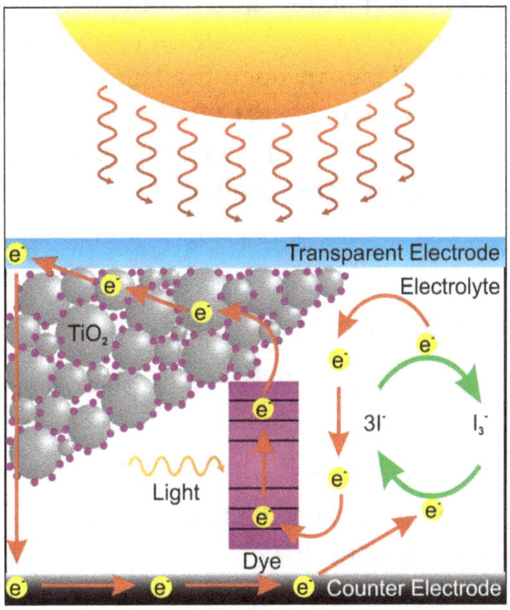

Operation of a Grätzel cell.

The dye molecules are quite small (nanometer sized), so in order to capture a reasonable amount of the incoming light the layer of dye molecules needs to be made fairly thick, much thicker than the molecules themselves. To address this problem, a nanomaterial is used as a scaffold to hold large numbers of the dye molecules in a 3-D matrix, increasing the number of molecules for any given surface area of cell. In existing designs, this scaffolding is provided by the semiconductor material, which serves double-duty.

Construction

In the case of the original Grätzel and O'Regan design, the cell has 3 primary parts. On top is a transparent anode made of fluoride-doped tin dioxide (SnO_2:F) deposited on the back of a (typically glass) plate. On the back of this conductive plate is a thin layer of titanium dioxide (TiO_2), which forms into a highly porous structure with an extremely high surface area. The (TiO_2) is chemically bound by a process called sintering. TiO_2 only absorbs a small fraction of the solar photons (those in the UV). The plate is then immersed in a mixture of a photosensitive ruthenium-polypyridine dye (also called molecular sensitizers) and a solvent. After soaking the film in the dye solution, a thin layer of the dye is left covalently bonded to the surface of the TiO_2. The bond is either an ester, chelating, or bidentate bridging linkage.

A separate plate is then made with a thin layer of the iodide electrolyte spread over a conductive sheet, typically platinum metal. The two plates are then joined and sealed together to prevent the electrolyte from leaking. The construction is simple enough that there are hobby kits available to hand-construct them. Although they use a number of "advanced" materials, these are inexpensive compared to the silicon needed for normal cells because they require no expensive manufacturing steps. TiO_2, for instance, is already widely used as a paint base.

One of the efficient DSSCs devices uses ruthenium-based molecular dye, e.g. [Ru(4,4'-dicarboxy-2,2'-bipyridine)$_2$(NCS)$_2$] (N3), that is bound to a photoanode via carboxylate moieties. The photoanode consists of 12 μm thick film of transparent 10–20 nm diameter TiO_2 nanoparticles covered with a 4 μm thick film of much larger (400 nm diameter) particles that scatter photons back into the transparent film. The excited dye rapidly injects an electron into the TiO_2 after light absorption. The injected electron diffuses through the sintered particle network to be collected at the front side transparent conducting oxide (TCO) electrode, while the dye is regenerated via reduction by a redox shuttle, I_3^-/I^-, dissolved in a solution. Diffusion of the oxidized form of the shuttle to the counter electrode completes the circuit.

Mechanism of DSSCs

The following steps convert in a DSSC photons (light) to current:

- The incident photon is absorbed by Ru complex photosensitizers adsorbed on the TiO_2 surface.
- The photosensitizers are excited from the ground state (S) to the excited state (S^*). The excited electrons are injected into the conduction band of the TiO_2 electrode. This results in the oxidation of the photosensitizer (S^+).

  $$S + h\nu \rightarrow S^*$$

  $$S^{\times} \xrightarrow{TiO_2} S^+ + e^-$$

- The injected electrons in the conduction band of TiO_2 are transported between TiO_2 nanoparticles with diffusion toward the back contact (TCO). And the electrons finally reach the counter electrode through the circuit.
- The oxidized photosensitizer (S^+) accepts electrons from the I^- ion redox mediator leading to regeneration of the ground state (S), and two I^--Ions are oxidized to elementary Iodine which reacts with I^- to the oxidized state, I_3^-.

  $$S^+ + e^- \rightarrow S$$

- The oxidized redox mediator, I_3^-, diffuses toward the counter electrode and then it is reduced to I^- ions.

  $$I_3^- + 2\,e^- \rightarrow 3\,I^-$$

The efficiency of a DSSC depends on four energy levels of the component: the excited state (approximately LUMO) and the ground state (HOMO) of the photosensitizer, the Fermi level of the TiO_2 electrode and the redox potential of the mediator (I^-/I_3^-) in the electrolyte.

Nanoplant-like Morphology

In DSSC, electrodes consisted of sintered semiconducting nanoparticles, mainly TiO_2 or ZnO. These nanoparticle DSSCs rely on trap-limited diffusion through the semiconductor nanoparticles for the electron transport. This limits the device efficiency since it is a slow transport mechanism. Recombination is more likely to occur at longer wavelengths of radiation. Moreover, sintering of nanoparticles requires a high temperature of about 450 °C, which restricts the fabrication of these cells to robust, rigid solid substrates. It has been proven that there is an increase in the efficiency of DSSC, if the sintered nanoparticle electrode is replaced by a specially designed electrode possessing an exotic 'nanoplant-like' morphology.

Operation

Sunlight enters the cell through the transparent SnO_2:F top contact, striking the dye on the surface of the TiO_2. Photons striking the dye with enough energy to be absorbed create an excited state of the dye, from which an electron can be "injected" directly into the conduction band of the TiO_2. From there it moves by diffusion (as a result of an electron concentration gradient) to the clear anode on top.

Meanwhile, the dye molecule has lost an electron and the molecule will decompose if another electron is not provided. The dye strips one from iodide in electrolyte below the TiO_2, oxidizing it into triiodide. This reaction occurs quite quickly compared to the time that it takes for the injected electron to recombine with the oxidized dye molecule, preventing this recombination reaction that would effectively short-circuit the solar cell.

The triiodide then recovers its missing electron by mechanically diffusing to the bottom of the cell, where the counter electrode re-introduces the electrons after flowing through the external circuit.

Efficiency

Several important measures are used to characterize solar cells. The most obvious is the total amount of electrical power produced for a given amount of solar power shining on the cell. Expressed as a percentage, this is known as the *solar conversion efficiency*. Electrical power is the product of current and voltage, so the maximum values for these measurements are important as well, J_{sc} and V_{oc} respectively. Finally, in order to understand the underlying physics, the "quantum efficiency" is used to compare the chance that one photon (of a particular energy) will create one electron.

In quantum efficiency terms, DSSCs are extremely efficient. Due to their "depth" in the nanostructure there is a very high chance that a photon will be absorbed, and the dyes are very effective at converting them to electrons. Most of the small losses that do exist in DSSC's are due to conduction losses in the TiO_2 and the clear electrode, or optical losses in the front electrode. The overall quantum efficiency for green light is about 90%, with the "lost" 10% being largely accounted for by the optical losses in the top electrode. The quantum efficiency of traditional designs vary, depending on their thickness, but are about the same as the DSSC.

In theory, the maximum voltage generated by such a cell is simply the difference between the (*quasi*-)Fermi level of the TiO_2 and the redox potential of the electrolyte, about 0.7 V under solar illumination conditions (V_{oc}). That is, if an illuminated DSSC is connected to a voltmeter in an "open circuit", it would read about 0.7 V. In terms of voltage, DSSCs offer slightly higher V_{oc} than silicon, about 0.7 V compared to 0.6 V. This is a fairly small difference, so real-world differences are dominated by current production, J_{sc}.

Although the dye is highly efficient at converting absorbed photons into free electrons in the TiO_2, only photons absorbed by the dye ultimately produce current. The rate of photon absorption depends upon the absorption spectrum of the sensitized TiO_2 layer and upon the solar flux spectrum. The overlap between these two spectra determines the maximum possible photocurrent. Typically used dye molecules generally have poorer absorption in the red part of the spectrum compared to silicon, which means that fewer of the photons in sunlight are usable for current generation. These factors limit the current generated by a DSSC, for comparison, a traditional silicon-based solar cell offers about 35 mA/cm^2, whereas current DSSCs offer about 20 mA/cm^2.

Overall peak power conversion efficiency for current DSSCs is about 11%. Current record for prototypes lies at 15%.

Degradation

DSSCs degrade when exposed to ultraviolet radiation. In 2014 air infiltration of the commonly-used amorphous Spiro-MeOTAD hole-transport layer was identified as the primary cause of the degradation, rather than oxidation. The damage could be avoided by the addition of an appropriate barrier.

The barrier layer may include UV stabilizers and/or UV absorbing luminescent chromophores (which emit at longer wavelengths which may be reabsorbed by the dye) and antioxidants to protect and improve the efficiency of the cell.

Advantages

DSSCs are currently the most efficient third-generation (2005 Basic Research Solar Energy Utilization 16) solar technology available. Other thin-film technologies are typically between 5% and 13%, and traditional low-cost commercial silicon panels operate between 14% and 17%. This makes DSSCs attractive as a replacement for existing technologies in "low density" applications like rooftop solar collectors, where the mechanical robustness and light weight of the glass-less collector is a major advantage. They may not be as attractive for large-scale deployments where higher-cost higher-efficiency cells are more viable, but even small increases in the DSSC conversion efficiency might make them suitable for some of these roles as well.

There is another area where DSSCs are particularly attractive. The process of injecting an electron directly into the TiO_2 is qualitatively different from that occurring in a traditional cell, where the electron is "promoted" within the original crystal. In theory, given low rates of production, the high-energy electron in the silicon could re-combine with its own hole, giving off a photon (or other form of energy) which does not result in current being generated. Although this particular case may not be common, it is fairly easy for an electron generated by another atom to combine with a hole left behind in a previous photoexcitation.

In comparison, the injection process used in the DSSC does not introduce a hole in the TiO_2, only an extra electron. Although it is energetically possible for the electron to recombine back into the dye, the rate at which this occurs is quite slow compared to the rate that the dye regains an electron from the surrounding electrolyte. Recombination directly from the TiO_2 to species in the electrolyte is also possible although, again, for optimized devices this reaction is rather slow. On the contrary, electron transfer from the platinum coated electrode to species in the electrolyte is necessarily very fast.

As a result of these favorable "differential kinetics", DSSCs work even in low-light conditions. DSSCs are therefore able to work under cloudy skies and non-direct sunlight, whereas traditional designs would suffer a "cutout" at some lower limit of illumination, when charge carrier mobility is low and recombination becomes a major issue. The cutoff is so low they are even being proposed for indoor use, collecting energy for small devices from the lights in the house.

A practical advantage which DSSCs share with most thin-film technologies, is that the cell's mechanical robustness indirectly leads to higher efficiencies at higher temperatures. In any semiconductor, increasing temperature will promote some electrons into the conduction band "mechanically". The fragility of traditional silicon cells requires them to be protected from the elements, typically by encasing them in a glass box similar to a greenhouse, with a metal backing for strength. Such systems suffer noticeable decreases in efficiency as the cells heat up internally. DSSCs are normally built with only a thin layer of conductive plastic on the front layer, allowing them to radiate away heat much easier, and therefore operate at lower internal temperatures.

Disadvantages

The major disadvantage to the DSSC design is the use of the liquid electrolyte, which has temperature stability problems. At low temperatures the electrolyte can freeze, halting power production and potentially leading to physical damage. Higher temperatures cause the liquid to expand, making sealing the panels a serious problem. Another disadvantage is that costly ruthenium (dye), platinum (catalyst) and conducting glass or plastic (contact) are needed to produce a DSSC. A third major drawback is that the electrolyte solution contains volatile organic compounds (or VOC's), solvents which must be carefully sealed as they are hazardous to human health and the environment. This, along with the fact that the solvents permeate plastics, has precluded large-scale outdoor application and integration into flexible structure.

Replacing the liquid electrolyte with a solid has been a major ongoing field of research. Recent experiments using solidified melted salts have shown some promise, but currently suffer from higher degradation during continued operation, and are not flexible.

Photocathodes and Tandem Cells

Dye sensitised solar cells operate as a photoanode (n-DSC), where photocurrent result from electron injection by the sensitized dye. Photocathodes (p-DSCs) operate in an inverse mode compared to the conventional n-DSC, where dye-excitation is followed by rapid electron transfer from a p-type semiconductor to the dye (dye-sensitized hole injection, instead of electron injection). Such p-DSCs and n-DSCs can be combined to construct tandem solar cells (pn-DSCs) and the theoretical efficiency of tandem DSCs is well beyond that of single-junction DSCs.

A standard tandem cell consists of one n-DSC and one p-DSC in a simple sandwich configuration with an intermediate electrolyte layer. n-DSC and p-DSC are connected in series, which implies that the resulting photocurrent will be controlled by the weakest photoelectrode, whereas photovoltages are additive. Thus, photocurrent matching is very important for the construction of highly efficient tandem pn-DSCs. However, unlike n-DSCs, fast charge recombination following dye-sensitized hole injection usually resulted in low photocurrents in p-DSC and thus hampered the efficiency of the overall device.

Researchers have found that using dyes comprising a perylenemonoimide (PMI) as the acceptor and an oligothiophene coupled to triphenylamine as the donor greatly improve the performance of p-DSC by reducing charge recombination rate following dye-sensitized hole injection. The researchers constructed a tandem DSC device with NiO on the p-DSC side and TiO_2 on the n-DSC side. Photocurrent matching was achieved through adjustment of NiO and TiO_2 film thicknesses to control the optical absorptions and therefore match the photocurrents of both electrodes. The energy conversion efficiency of the device is 1.91%, which exceeds the efficiency of its individual components, but is still much lower than that of high performance n-DSC devices (6%–11%). The results are still promising since the tandem DSC was in itself rudimentary. The dramatic improvement in performance in p-DSC can eventually lead to tandem devices with much greater efficiency than lone n-DSCs.

Using electron transporting materials such as PCBM, TiO_2 and ZnO instead of the conventional liquid redox couple electrolyte, researchers have managed to fabricate the solid state p-DSC (p-ssDSCs), aiming for solid state tandem dye sensitized solar cells.

Development

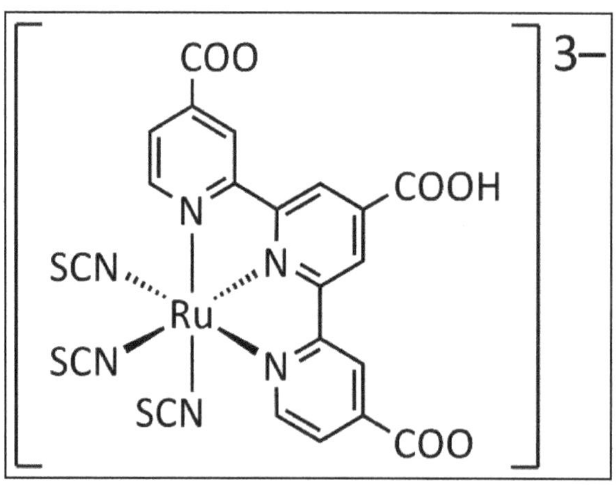

"Black Dye", an anionic Ru-terpyridine complex.

The dyes used in early experimental cells (circa 1995) were sensitive only in the high-frequency end of the solar spectrum, in the UV and blue. Newer versions were quickly introduced (circa 1999) that had much wider frequency response, notably "triscarboxy-ruthenium terpyridine" [Ru(4,4',4"-$(COOH)_3$-terpy)$(NCS)_3$], which is efficient right into the low-frequency range of red and IR light. The wide spectral response results in the dye having a deep brown-black color, and is referred to simply as "black dye". The dyes have an excellent chance of converting a photon into an electron, originally around 80% but improving to almost perfect conversion in more recent dyes, the overall efficiency is about 90%, with the "lost" 10% being largely accounted for by the optical losses in top electrode.

A solar cell must be capable of producing electricity for at least twenty years, without a significant decrease in efficiency (life span). The "black dye" system was subjected to 50 million cycles, the equivalent of ten years' exposure to the sun in Switzerland. No discernible performance decrease was observed. However the dye is subject to breakdown in high-light situations. Over the last decade an extensive research program has been carried out to address these concerns. The newer dyes included 1-ethyl-3 methylimidazolium tetrocyanoborate [EMIB$(CN)_4$] which is extremely light- and temperature-stable, copper-diselenium [Cu(In,GA)Se_2] which offers higher conversion efficiencies, and others with varying special-purpose properties.

DSSCs are still at the start of their development cycle. Efficiency gains are possible and have recently started more widespread study. These include the use of quantum dots for conversion of higher-energy (higher frequency) light into multiple electrons, using solid-state electrolytes for better temperature response, and changing the doping of the TiO_2 to better match it with the electrolyte being used.

New Developments

A group of researchers at the Swiss Federal Institute of Technology has reportedly increased the thermostability of DSC by using amphiphilic ruthenium sensitizer in conjunction with quasi-solid-state gel electrolyte. The stability of the device matches that of a conventional inorganic silicon-based solar cell. The cell sustained heating for 1,000 h at 80 °C.

The group has previously prepared a ruthenium amphiphilic dye Z-907 (cis-Ru(H_2dcbpy)(dnbpy)$(NCS)_2$, where the ligand H_2dcbpy is 4,4′-dicarboxylic acid-2,2′-bipyridine and dnbpy is 4,4′-dinonyl-2,2′-bipyridine) to increase dye tolerance to water in the electrolytes. In addition, the group also prepared a quasi-solid-state gel electrolyte with a 3-methoxypropionitrile (MPN)-based liquid electrolyte that was solidified by a photochemically stable fluorine polymer, polyvinylidenefluoride-co-hexafluoropropylene (PVDF-HFP).

The use of the amphiphilic Z-907 dye in conjunction with the polymer gel electrolyte in DSC achieved an energy conversion efficiency of 6.1%. More importantly, the device was stable under thermal stress and soaking with light. The high conversion efficiency of the cell was sustained after heating for 1,000 h at 80 °C, maintaining 94% of its initial value. After accelerated testing in a solar simulator for 1,000 h of light-soaking at 55 °C (100 mW cm^{-2}) the efficiency had decreased by less than 5% for cells covered with an ultraviolet absorbing polymer film. These results are well within the limit for that of traditional inorganic silicon solar cells.

The enhanced performance may arise from a decrease in solvent permeation across the sealant due to the application of the polymer gel electrolyte. The polymer gel electrolyte is quasi-solid at room temperature, and becomes a viscous liquid (viscosity: 4.34 mPa·s) at 80 °C compared with the traditional liquid electrolyte (viscosity: 0.91 mPa·s). The much improved stabilities of the device under both thermal stress and soaking with light has never before been seen in DSCs, and they match the durability criteria applied to solar cells for outdoor use, which makes these devices viable for practical application.

The first successful solid-hybrid dye-sensitized solar cells were reported.

To improve electron transport in these solar cells, while maintaining the high surface area needed for dye adsorption, two researchers have designed alternate semiconductor morphologies, such as arrays of nanowires and a combination of nanowires and nanoparticles, to provide a direct path to the electrode via the semiconductor conduction band. Such structures may provide a means to improve the quantum efficiency of DSSCs in the red region of the spectrum, where their performance is currently limited.

On August 2006, to prove the chemical and thermal robustness of the 1-ethyl-3 methylimidazolium tetracyanoborate solar cell, the researchers subjected the devices to heating at 80 °C in the dark for 1000 hours, followed by light soaking at 60 °C for 1000 hours. After dark heating and light soaking, 90% of the initial photovoltaic efficiency was maintained – the first time such excellent thermal stability has been observed for a liquid electrolyte that exhibits such a high conversion efficiency. Contrary to silicon solar cells, whose performance declines with increasing temperature, the dye-sensitized solar-cell devices were only negligibly influenced when increasing the operating temperature from ambient to 60 °C.

Wayne Campbell at Massey University, New Zealand, has experimented with a wide variety of organic dyes based on porphyrin. In nature, porphyrin is the basic building block of the hemoproteins, which include chlorophyll in plants and hemoglobin in animals. He reports efficiency on the order of 5.6% using these low-cost dyes.

Nature Materials demonstrated cell efficiencies of 8.2% using a new solvent-free liquid redox electrolyte consisting of a melt of three salts, as an alternative to using organic solvents as an electrolyte solution. Although the efficiency with this electrolyte is less than the 11% being delivered using the existing iodine-based solutions, the team is confident the efficiency can be improved.

A group of researchers at Georgia Tech made dye-sensitized solar cells with a higher effective surface area by wrapping the cells around a quartz optical fiber. The researchers removed the cladding from optical fibers, grew zinc oxide nanowires along the surface, treated them with dye molecules, surrounded the fibers by an electrolyte and a metal film that carries electrons off the fiber. The cells are six times more efficient than a zinc oxide cell with the same surface area. Photons bounce inside the fiber as they travel, so there are more chances to interact with the solar cell and produce more current. These devices only collect light at the tips, but future fiber cells could be made to absorb light along the entire length of the fiber, which would require a coating that is conductive as well as transparent. Max Shtein of the University of Michigan said a sun-tracking system would not be necessary for such cells, and would work on cloudy days when light is diffuse.

Researchers at the École Polytechnique Fédérale de Lausanne and at the Université du Québec à Montréal claim to have overcome two of the DSC's major issues:

- "New molecules" have been created for the electrolyte, resulting in a liquid or gel that is transparent and non-corrosive, which can increase the photovoltage and improve the cell's output and stability.
- At the cathode, platinum was replaced by cobalt sulfide, which is far less expensive, more efficient, more stable and easier to produce in the laboratory.

Dyesol and Tata Steel Europe announced in June the development of the world's largest dye sensitized photovoltaic module, printed onto steel in a continuous line.

Dyesol and CSIRO announced in October a Successful Completion of Second Milestone in Joint Dyesol / CSIRO Project. Dyesol Director Gordon Thompson said, "The materials developed during this joint collaboration have the potential to significantly advance the commercialisation of DSC in a range of applications where performance and stability are essential requirements. Dyesol is extremely encouraged by the breakthroughs in the chemistry allowing the production of the target molecules. This creates a path to the immediate commercial utilisation of these new materials."

Dyesol and Tata Steel Europe announced in November the targeted development of Grid Parity Competitive BIPV solar steel that does not require government subsidised feed in tariffs. TATA-Dyesol "Solar Steel" Roofing is currently being installed on the Sustainable Building Envelope Centre (SBEC) in Shotton, Wales.

Northwestern University researchers announced a solution to a primary problem of DSSCs, that of difficulties in using and containing the liquid electrolyte and the consequent relatively short useful life of the device. This is achieved through the use of nanotechnology and the conversion of the liquid electrolyte to a solid. The current efficiency is about half that of silicon cells, but the cells are lightweight and potentially of much lower cost to produce.

During the last 5–10 years, a new kind of DSSC has been developed – the solid state dye-sensitized solar cell. In this case the liquid electrolyte is replaced by one of several solid hole conducting materials. From 2009 to 2013 the efficiency of Solid State DSSCs has dramatically increased from 4% to 15%. Michael Grätzel announced the fabrication of Solid State DSSCs with 15.0% efficiency, reached by the means of a hybrid perovskite $CH_3NH_3PbI_3$ dye, subsequently deposited from the separated solutions of CH_3NH_3I and PbI_2.

The first architectural integration was demonstrated at EPFL's new convention center in partnership with Romande Energie. The total surface is 300 m², in 1400 modules of 50 cm x 35 cm. Designed by artists Daniel Schlaepfer and Catherine Bolle.

Researchers have investigated the role of surface plasmon resonances present on gold nanorods in the performance of dye-sensitized solar cells. They found that with an increase nanorod concentration, the light absorption grew linearly; however, charge extraction was also dependent on the concentration. With an optimized concentration, they found that the overall power conversion efficiency improved from 5.31 to 8.86% for Y123 dye-sensitized solar cells.

The synthesis of one-dimensional TiO_2 nanostructures directly on fluorine-doped tin oxide glass substrates was successful demonstrated via a two-stop solvothermal reaction. Additionally, through a TiO_2 sol treatment, the performance of the dual TiO_2 nanowire cells was enhanced, reaching a power conversion efficiency of 7.65%.

Stainless steel based counter-electrodes for DSSCs have been reported which further reduce cost compared to conventional platinum based counter electrode and are suitable for outdoor application.

Researchers from EPFL have advanced the DSSCs based on copper complexes redox electrolytes, which have achieved 13.1% efficiency under standard AM1.5G, 100 mW/cm² conditions and record 32% efficiency under 1000 lux of indoor light.

Researchers from Uppsala University have used n-type semiconductors instead of redox electrolyte to fabricate solid state p-type dye sensitized solar cells.

Market Introduction

Several commercial providers are promising availability of DSCs in the near future:

- Dyesol officially opened its new manufacturing facilities in Queanbeyan Australia on 7 October 2008. It has subsequently announced partnerships with Tata Steel (TATA-Dyesol) and Pilkington Glass (Dyetec-Solar) for the development and large scale manufacture of DSC BIPV. Dyesol has also entered working relationships with Merck, Umicore, CSIRO, Japanese Ministry of Economy and Trade, Singapore Aerospace Manufacturing and a joint Venture with TIMO Korea (Dyesol-TIMO).
- Solaronix, a Swiss company specialized in the production of DSC materials since 1993, has extended their premises in 2010 to host a manufacturing pilot line of DSC modules.
- SolarPrint was founded in Ireland in 2008 by Dr. Mazhar Bari, Andre Fernon and Roy Horgan. SolarPrint was the first Ireland-based commercial entity involved in the manufacturing of PV technology. SolarPrint's innovation was the solution to the solvent-based electrolyte which to date has prohibited the mass commercialisation of DSSC. The company went into receivership in 2014 and was wound up.
- G24innovations founded in 2006, based in Cardiff, South Wales, UK. On 17 October 2007, claimed the production of the first commercial grade dye sensitised thin films.
- Sony Corporation has developed dye-sensitized solar cells with an energy conversion efficiency of 10%, a level seen as necessary for commercial use.
- Tasnee Enters Strategic Investment Agreement with Dyesol.
- H.Glass was founded 2011 in Switzerland. H.Glass has put enormous efforts to create industrial process for the DSSC technologie - the first results where shown at the EXPO 2015 in Milano at the Austrian Pavilion. The milestone for DSSC is the Science Tower in Austria - it is the largest installation of DSSC in the world - carried out by SFL technologies.
- Exeger Operations AB, Sweden, has built a factory in Stockholm with a capacity of 300,000m2. SoftBank Group Corp. has made two investments of US$10M in Exeger during 2019.

SOLAR PANEL

Solar panels (also known as "PV panels") are used to convert light from the sun, which is composed of particles of energy called "photons", into electricity that can be used to power electrical loads.

Solar panels can be used for a wide variety of applications including remote power systems for cabins, telecommunications equipment, remote sensing, and of course for the production of electricity by residential and commercial solar electric systems.

Working of Solar Panels

Solar panels collect clean renewable energy in the form of sunlight and convert that light into electricity which can then be used to provide power for electrical loads. Solar panels are comprised of several individual solar cells which are themselves composed of layers of silicon, phosphorous (which provides the negative charge), and boron (which provides the positive charge). Solar panels absorb the photons and in doing so initiate an electric current. The resulting energy generated from photons striking the surface of the solar panel allows electrons to be knocked out of their atomic orbits and released into the electric field generated by the solar cells which then pull these free electrons into a directional current. This entire process is known as the Photovoltaic Effect. An average home has more than enough roof area for the necessary number of solar panels to produce enough solar electricrity to supply all of its power needs excess electricity generated goes onto the main power grid, paying off in electricity use at night.

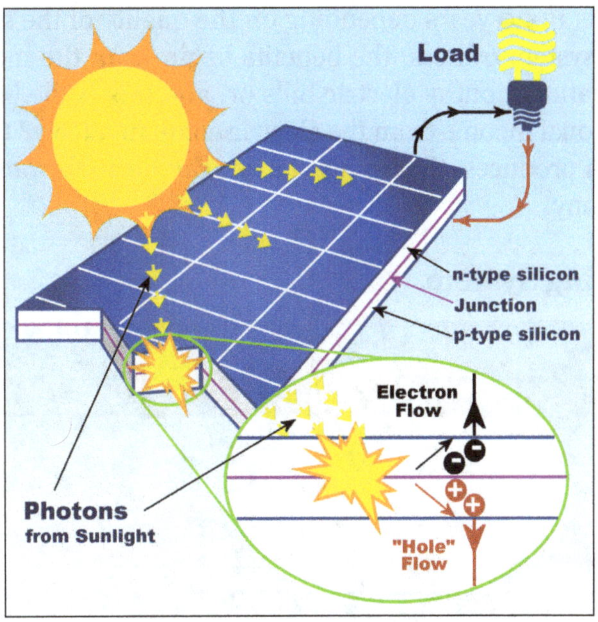

In a well-balanced grid-connected configuration, a solar array generates power during the day that is then used in the home at night. Net metering programs allow solar generator owners to get paid if their system produces more power than what is needed in the home. In off-grid solar applications, a battery bank, charge controller, and in most cases, an inverter are necessary components. The solar array sends direct current (DC) electricity through the charge controller to the battery bank. The power is then drawn from the battery bank to the inverter, which converts the DC current into alternating current (AC) that can be used for non-DC appliances. Assisted by an inverter, solar panel arrays can be sized to meet the most demanding electrical load requirements. The AC current can be used to power loads in homes or commercial buildings, recreational vehicles and boats, remote cabins, cottages, or homes, remote traffic controls, telecommunications equipment, oil and gas flow monitoring, RTU, SCADA, and much more.

Benefits of Solar Panels

Using solar panels is a very practical way to produce electricity for many applications. The obvious would have to be off-grid living. Living off-grid means living in a location that is not serviced by the main electric utility grid. Remote homes and cabins benefit nicely from solar power systems. No longer is it necessary to pay huge fees for the installation of electric utility poles and cabling from the nearest main grid access point. A solar electric system is potentially less expensive and can provide power for upwards of three decades if properly maintained.

Besides the fact that solar panels make it possible to live off-grid, perhaps the greatest benefit that you would enjoy from the use of solar power is that it is both a clean and a renewable source of energy. With the advent of global climate change, it has become more important that we do whatever we can to reduce the pressure on our atmosphere from the emission of greenhouse gases. Solar panels have no moving parts and require little maintenance. They are ruggedly built and last for decades when porperly maintained.

Last, but not least, of the benefits of solar panels and solar power is that, once a system has paid for its initial installation costs, the electricity it produces for the remainder of the system's lifespan, which could be as much as 15-20 years depending on the quality of the system, is absolutely free! For grid-tie solar power system owners, the benefits begin from the moment the system comes online, potentially eliminating monthy electric bills or, and this is the best part, actually earning the system's owner additional income from the electric company. How? If you use less power than your solar electric system produces, that excess power can be sold, sometimes at a premium, to your electric utility company.

Photovoltaic Mounting System

Solar panel mounting system on roof of Pacifica wastewater treatment plant.

Photovoltaic mounting systems (also called solar module racking) are used to fix solar panels on surfaces like roofs, building facades, or the ground. These mounting systems generally enable retrofitting of solar panels on roofs or as part of the structure of the building (called BIPV).

Roof Mounting

PV panels mounted on roof.

Workers install residential rooftop solar panels.

The solar array of a PV system can be mounted on rooftops, generally with a few inches gap and parallel to the surface of the roof. If the rooftop is horizontal, the array is mounted with each panel aligned at an angle. If the panels are planned to be mounted before the construction of the roof, the roof can be designed accordingly by installing support brackets for the panels before the materials for the roof are installed. The installation of the solar panels can be undertaken by the crew responsible for installing the roof. If the roof is already constructed, it is relatively easy to retrofit panels directly on top of existing roofing structures. For a small minority of roofs (often not built to code) that are designed so that it is capable of bearing only the weight of the roof, installing solar panels demands that the roof structure must be strengthened beforehand. In all cases of retrofits particular consideration to weather sealing is necessary There are many low-weight designs for PV systems that can be used on either sloped or flat roofs (e.g. plastic wedges or the PV-pod), most however, rely on a type of extruded aluminum rails (e.g. Unirac). Recently, tension-based PV racking solutions have been tested successfully that reduce weight and cost. In some cases, converting to composition shingles, the weight of the removed roof materials can compensate the additional weight of the panels structure. The general practice for installation of roof-mounted solar panels include having a support bracket per hundred watts of panels.

Ground-mounted

Ground-mounted PV systems are usually large, utility-scale photovoltaic power stations. The PV array consist of solar modules held in place by racks or frames that are attached to ground-based mounting supports.

Ground-based mounting supports include:

- Pole mounts, which are driven directly into the ground or embedded in concrete.
- Foundation mounts, such as concrete slabs or poured footings.
- Ballasted footing mounts, such as concrete or steel bases that use weight to secure the solar module system in position and do not require ground penetration. This type of mounting

system is well suited for sites where excavation is not possible such as capped landfills and simplifies decommissioning or relocation of solar module systems.

Mounting as a Shade Structure

PV panels as external shading device in zero-energy building, Singapore.

Solar panels can also be mounted as shade structures where the solar panels can provide shade instead of patio covers. The cost of such shading systems are generally different from standard patio covers, especially in cases where the entire shade required is provided by the panels. The support structure for the shading systems can be normal systems as the weight of a standard PV array is between 3 and 5 pounds/ft^2. If the panels are mounted at an angle steeper than normal patio covers, the support structures may require additional strengthening. Other issues that are considered include:

- Simplified array access for maintenance.
- Module wiring may be concealed to maintain the aesthetics of the shading structure.
- Growing vines around the structure must be avoided as they may come in contact with the wiring.

Building-integrated Photovoltaics

Building-integrated photovoltaics (BIPV) are photovoltaic materials that are used to replace conventional building materials in parts of the building envelope such as the roof (tiles), skylights, or facades. They are increasingly being incorporated into the construction of new buildings as a principal or ancillary source of electrical power, although existing buildings may be retrofitted with BIPV modules as well. The advantage of integrated photovoltaics over more common non-integrated systems is that the initial cost can be offset by reducing the amount spent on building materials and labor that would normally be used to construct the part of the building that the BIPV modules replace.

The CIS Tower in Manchester, England was clad in PV panels at a cost of £5.5 million. It started feeding electricity to the National Grid in November 2005.

Building-adapted photovoltaics (BAPV) uses solar modules to create solar PV windows and this way, also to retrofit existing building.

Orientation and Inclination

A solar cell performs the best when its surface is perpendicular to the sun's rays, which change continuously over the course of the day and season. It is a common practice to tilt a fixed PV module (without solar tracker) at the same angle as the latitude of array's location to maximize the annual energy yield of module. For example, rooftop PV module at the tropics provides highest annual energy yield when inclination of panel surface is close to horizontal direction. A study in the tropics showed that the orientation of low-slope rooftop PV has negligible impact on annual energy yield, but in the case of PV external sunshade applications, east facade and panel slope of 30–40° are the most suitable location and inclination.

SOLAR TRACKER

A solar tracker is a device that orients a payload toward the Sun. Payloads are usually solar panels, parabolic troughs, fresnel reflectors, lenses or the mirrors of a heliostat.

For flat-panel photovoltaic systems, trackers are used to minimize the angle of incidence between the incoming sunlight and a photovoltaic panel, sometimes known as the cosine error. Reducing this angle increases the amount of energy produced from a fixed amount of installed power generating capacity. In standard photovoltaic applications, it was predicted in 2008-2009 that trackers could be used in at least 85% of commercial installations greater than one megawatt from 2009 to 2012.

Largest single axis tracker project in Asia- 172MW- Arctech Solar.

8-megawatt PV plant using horizontal single axis tracker, Greece.

As the pricing, reliability and performance of single-axis trackers have improved, the systems have been installed in an increasing percentage of utility-scale projects. According to data from Wood-Mackenzie/GTM Research, global solar tracker shipments hit a record 14.5 gigawatts in 2017. This represents growth of 32 percent year-over-year, with similar or greater growth projected as large-scale solar deployment accelerates.

Suntactics dual axis solar trackers are used for small for medium sized solar production farms. Useful for small business solar power and battery charging.

In concentrator photovoltaics (CPV) and concentrated solar power (CSP) applications, trackers are used to enable the optical components in the CPV and CSP systems. The optics in concentrated solar applications accept the direct component of sunlight light and therefore must be oriented appropriately to collect energy. Tracking systems are found in all concentrator applications because such systems collect the sun's energy with maximum efficiency when the optical axis is aligned with incident solar radiation.

Sunlight has two components, the "direct beam" that carries about 90% of the solar energy, and the "diffuse sunlight" that carries the remainder – the diffuse portion is the blue sky on a clear day, and is a larger proportion of the total on cloudy days. As the majority of the energy is in the direct beam, maximizing collection requires the Sun to be visible to the panels for as long as possible. However, on cloudier days the ratio of direct vs. diffuse light can be as low as 60:40 or even lower.

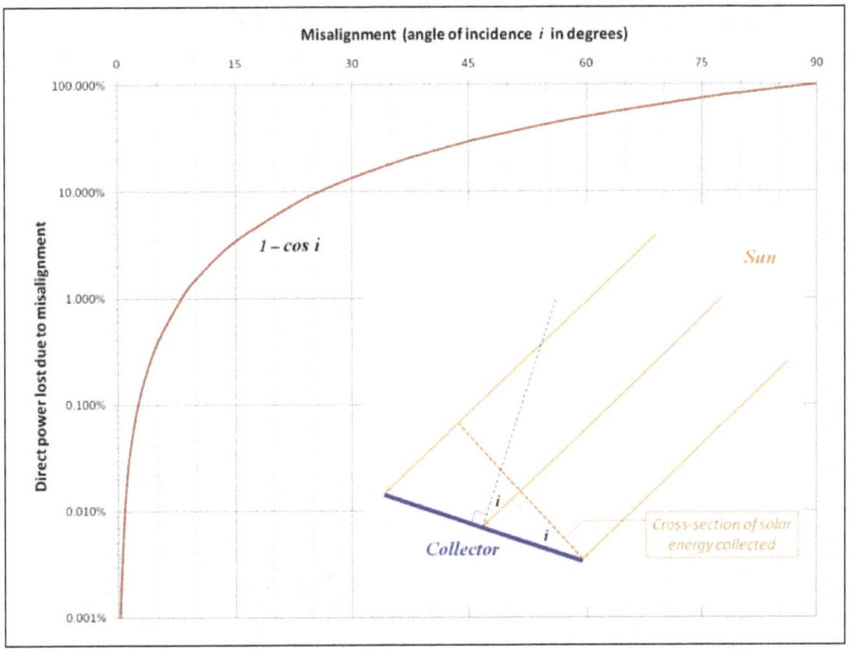

The effective collection area of a flat-panel solar collector varies with the cosine of the misalignment of the panel with the Sun.

The energy contributed by the direct beam drops off with the cosine of the angle between the incoming light and the panel. In addition, the reflectance (averaged across all polarizations) is approximately constant for angles of incidence up to around 50°, beyond which reflectance degrades rapidly.

Direct power lost (%) due to misalignment (angle i) where Lost = 1 - cos(i)				
i	Lost	i	hours	Lost
0°	0%	15°	1	3.4%
1°	0.015%	30°	2	13.4%
3°	0.14%	45°	3	30%
8°	1%	60°	4	>50%
23.4°	8.3%	75°	5	>75%

For example, trackers that have accuracies of ± 5° can deliver greater than 99.6% of the energy delivered by the direct beam plus 100% of the diffuse light. As a result, high accuracy tracking is not typically used in non-concentrating PV applications.

The purpose of a tracking mechanism is to follow the Sun as it moves across the sky.The complex path of the Sun is simplified by considering its daily east-west motion separately from its yearly north-south variation with the seasons of the year.

Solar Energy Intercepted

The amount of solar energy available for collection from the direct beam is the amount of light intercepted by the panel. This is given by the area of the panel multiplied by the cosine of the angle of incidence of the direct beam. Or put another way, the energy intercepted is equivalent to the area of the shadow cast by the panel onto a surface perpendicular to the direct beam.

This cosine relationship is very closely related to the observation formalized in 1760 by Lambert's cosine law. This describes that the observed brightness of an object is proportional to the cosine of the angle of incidence of the light illuminating it.

Reflective Losses

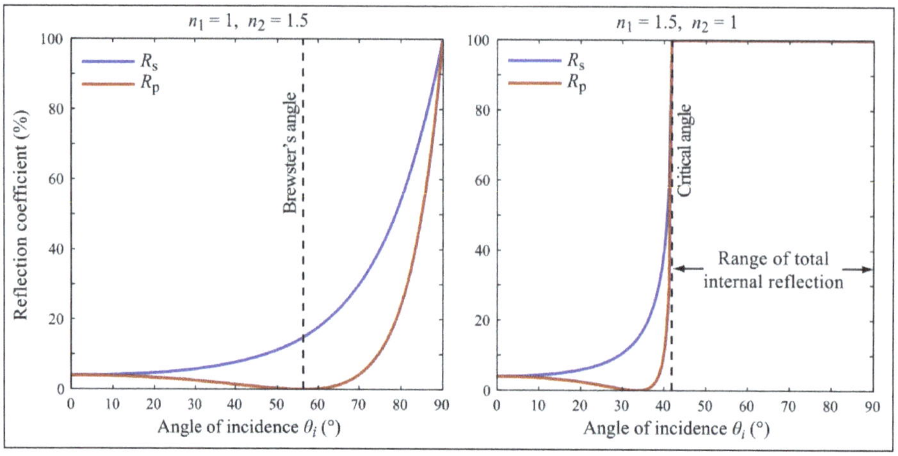

Variation of reflectance with angle of incidence.

Not all of the light intercepted is transmitted into the panel - a little is reflected at its surface. The amount reflected is influenced by both the refractive index of the surface material and the angle of incidence of the incoming light. The amount reflected also differs depending on the polarization of the incoming light. Incoming sunlight is a mixture of all polarizations. Averaged over all polarizations, the reflective losses are approximately constant up to angles of incidence up to around 50° beyond which it degrades rapidly.

Daily East-west Motion of the Sun

The Sun travels through 360 degrees east to west per day, but from the perspective of any fixed location the visible portion is 180 degrees during an average 1/2 day period (more in spring and summer; less, in fall and winter). Local horizon effects reduce this somewhat, making the effective motion about 150 degrees. A solar panel in a fixed orientation between the dawn and sunset extremes will see a motion of 75 degrees to either side, and thus, will lose over 75% of the energy in the morning and evening. Rotating the panels to the east and west can help recapture those losses. A tracker that only attempts to compensate for the east-west movement of the Sun is known as a single-axis tracker.

Seasonal North-south Motion of the Sun

Due to the tilt of the Earth's axis, the Sun also moves through 46 degrees north and south during a year. The same set of panels set at the midpoint between the two local extremes will thus see the Sun move 23 degrees on either side. Thus, an optimally aligned single-axis tracker will only lose 8.3% at the summer and winter seasonal extremes, or around 5% averaged over a year. Conversely a vertically or horizontally aligned single-axis tracker will lose considerably more as a result of these seasonal variations in the Sun's path. For example, a vertical tracker at a site at 60° latitude will lose up to 40% of the available energy in summer, while a horizontal tracker located at 25° latitude will lose up to 33% in winter.

A tracker that accounts for both the daily and seasonal motions is known as a dual-axis tracker. Generally speaking, the losses due to seasonal angle changes is complicated by changes in the length of the day, increasing collection in the summer in northern or southern latitudes. This biases collection toward the summer, so if the panels are tilted closer to the average summer angles, the total yearly losses are reduced compared to a system tilted at the spring/fall equinox angle (which is the same as the site's latitude).

There is considerable argument within the industry whether the small difference in yearly collection between single and dual-axis trackers makes the added complexity of a two-axis tracker worthwhile. A recent review of actual production statistics from southern Ontario suggested the difference was about 4% in total, which was far less than the added costs of the dual-axis systems. This compares unfavourably with the 24-32% improvement between a fixed-array and single-axis tracker.

Other Factors

Clouds

The above models assume uniform likelihood of cloud cover at different times of day or year. In different climate zones cloud cover can vary with seasons, affecting the averaged performance figures Alternatively, for example in an area where cloud cover on average builds up during the day, there can be particular benefits in collecting morning sun.

Atmosphere

The distance that sunlight has to travel through the atmosphere increases as the sun approaches the horizon, as the sunlight has to travel diagonally through the atmosphere. As the path length through the atmosphere increases, the solar intensity reaching the collector decreases. This increasing path length is referred to as the air mass (AM) or air mass coefficient, where AM0 is at the top of the atmosphere, AM1 refers to the direct vertical path down to sea-level with Sun overhead, and AM greater than 1 refers to diagonal paths as the Sun approaches the horizon.

Even though the sun may not feel particularly hot in the early mornings or during the winter months, the diagonal path through the atmosphere has a less than expected impact on the solar intensity. Even when the sun is only 15° above the horizon the solar intensity can be around 60% of its maximum value, around 50% at 10° and 25% at only 5° above the horizon. Therefore, trackers can deliver benefit by collecting the significant energy available when the Sun is close to the horizontal is this is possible.

Solar Cell Efficiency

Of course the underlying power conversion efficiency of a photovoltaic cell has a major influence on the end result, regardless of whether tracking is employed or not. Of particular relevance to the benefits of tracking are the following:

Molecular Structure

Much research is aimed at developing surface materials to guide the maximum amount of energy down into the cell and minimize reflective losses.

Temperature

Photovoltaic solar cell efficiency decreases with increasing temperature, at the rate of about 0.4%/°C. For example, 20% higher efficiency at 10 °C in early morning or winter as compared with 60 °C in the heat of the day or summer. Therefore, trackers can deliver additional benefit by collecting early morning and winter energy when the cells are operating at their highest efficiency.

Types of Solar Collector

Solar collectors may be:

- Non-concentrating flat-panels, usually photovoltaic or hot-water,
- Concentrating systems, of a variety of types.

Solar collector mounting systems may be fixed (manually aligned) or tracking. Different types of solar collector and their location (latitude) require different types of tracking mechanism. Tracking systems may be configured as:

- Fixed collector/moving mirror - i.e. *Heliostat.*
- Moving collector.

Non-tracking Fixed Mount

Residential and small-capacity commercial or industrial rooftop solar panels and solar water heater panels are usually fixed, often flush-mounted on an appropriately facing pitched roof. Advantages of fixed mounts over trackers include the following:

- Mechanical Advantages: Simple to manufacture, lower installation and maintenance costs.
- Wind-loading: It is easier and cheaper to provision a sturdy mount; all mounts other than fixed flush-mounted panels must be carefully designed having regard to wind loading due to greater exposure.
- Indirect light: Approximately 10% of the incident solar radiation is diffuse light, available at any angle of misalignment with the Sun.
- Tolerance to misalignment: Effective collection area for a flat-panel is relatively insensitive to quite high levels of misalignment with the Sun – for example even a 25° misalignment reduces the direct solar energy collected by less than 10%.

Fixed mounts are usually used in conjunction with non-concentrating systems, however an important class of non-tracking concentrating collectors, of particular value in the 3rd world, are portable solar cookers. These utilize relatively low levels of concentration, typically around 2 to 8 Suns and are manually aligned.

Trackers

Even though a fixed flat-panel can be set to collect a high proportion of available noon-time energy, significant power is also available in the early mornings and late afternoons when the misalignment with a fixed panel becomes excessive to collect a reasonable proportion of the available energy. For example, even when the Sun is only 10° above the horizon the available energy can be around half the noon-time energy levels (or even greater depending on latitude, season, and atmospheric conditions).

Thus the primary benefit of a tracking system is to collect solar energy for the longest period of the day, and with the most accurate alignment as the Sun's position shifts with the seasons.

In addition, the greater the level of concentration employed, the more important accurate tracking becomes, because the proportion of energy derived from direct radiation is higher, and the region where that concentrated energy is focused becomes smaller.

Fixed Collector/Moving Mirror

Many collectors cannot be moved, for example high-temperature collectors where the energy is recovered as hot liquid or gas (e.g. steam). Other examples include direct heating and lighting of buildings and fixed in-built solar cookers, such as Scheffler reflectors. In such cases it is necessary to employ a moving mirror so that, regardless of where the Sun is positioned in the sky, the Sun's rays are redirected onto the collector.

Due to the complicated motion of the Sun across the sky, and the level of precision required to correctly aim the Sun's rays onto the target, a heliostat mirror generally employs a dual axis tracking system, with at least one axis mechanized. In different applications, mirrors may be flat or concave.

Moving Collector

Trackers can be grouped into classes by the number and orientation of the tracker's axes. Compared to a fixed mount, a single axis tracker increases annual output by approximately 30%, and a dual axis tracker an additional 10-20%.

Photovoltaic trackers can be classified into two types: standard photovoltaic (PV) trackers and concentrated photovoltaic (CPV) trackers. Each of these tracker types can be further categorized by the number and orientation of their axes, their actuation architecture and drive type, their intended applications, their vertical supports and foundation.

Floating Mount

Floating islands of solar panels are being installed on reservoirs and lakes in the Netherlands, China, the UK and Japan. The sun-tracking system controlling the direction of the panels operates automatically according to the time of year, changing position by means of ropes attached to buoys.

Floating Ground Mount

Solar trackers can be built using a "floating" foundation, which sits on the ground without the need for invasive concrete foundations. Instead of placing the tracker on concrete foundations, the tracker is placed on a gravel pan that can be filled with a variety of materials, such as sand or gravel, to secure the tracker to the ground. These "floating" trackers can sustain the same wind load as a traditional fixed mounted tracker. The use of floating trackers increases the number of potential sites for commercial solar projects since they can be placed on top of capped landfills or in areas where excavated foundations are not feasible.

Motion Free Optical Tracking

Solar trackers can be built without the need for mechanical tracking equipment. These are called motion free optical tracking, there has been some series of advancements in this technology over the past few decades.

Non-concentrating Photovoltaic (PV) Trackers

Photovoltaic panels accept both direct and diffuse light from the sky. The panels on standard photovoltaic trackers gather both the available direct and diffuse light. The tracking functionality in standard photovoltaic trackers is used to minimize the angle of incidence between incoming light and the photovoltaic panel. This increases the amount of energy gathered from the direct component of the incoming sunlight.

The physics behind standard photovoltaic (PV) trackers works with all standard photovoltaic module technologies. These include all types of crystalline silicon panels (either mono-Si, or multi-Si) and all types of thin film panels (amorphous silicon, CdTe, CIGS, microcrystalline).

Concentrator Photovoltaic (CPV) Trackers

3-megawatt CPV plant using dual axis trackers in Golmud, China.

200-kilowatt CPV modules on dual axis tracker in Qingdao, China.

The optics in CPV modules accept the direct component of the incoming light and therefore must be oriented appropriately to maximize the energy collected. In low concentration applications a portion of the diffuse light from the sky can also be captured. The tracking functionality in CPV modules is used to orient the optics such that the incoming light is focused to a photovoltaic collector.

CPV modules that concentrate in one dimension must be tracked normal to the Sun in one axis. CPV modules that concentrate in two dimensions must be tracked normal to the Sun in two axes.

Accuracy Requirements

The physics behind CPV optics requires that tracking accuracy increase as the systems concentration ratio increases. However, for a given concentration, nonimaging optics provide the widest possible acceptance angles, which may be used to reduce tracking accuracy.

In typical high concentration systems tracking accuracy must be in the ± 0.1° range to deliver approximately 90% of the rated power output. In low concentration systems, tracking accuracy must be in the ± 2.0° range to deliver 90% of the rated power output. As a result, high accuracy tracking systems are typical.

Technologies Supported

Concentrated photovoltaic trackers are used with refractive and reflective based concentrator systems. There are a range of emerging photovoltaic cell technologies used in these systems. These range from conventional, crystalline silicon-based photovoltaic receivers to germanium-based triple junction receivers.

Single Axis Trackers

Single axis trackers have one degree of freedom that acts as an axis of rotation. The axis of rotation of single axis trackers is typically aligned along a true North meridian. It is possible to align them in any cardinal direction with advanced tracking algorithms. There are several common implementations of single axis trackers. These include horizontal single axis trackers (HSAT), horizontal single axis tracker with tilted modules (HTSAT), vertical single axis trackers (VSAT), tilted single axis trackers (TSAT) and polar aligned single axis trackers (PSAT). The orientation of the module with respect to the tracker axis is important when modeling performance.

Horizontal

Horizontal Single Axis Tracker (HSAT)

4MW horizontal single axis tracker
in Vellakoil, Tamil Nadu, India.

The axis of rotation for horizontal single axis tracker is horizontal with respect to the ground. The posts at either end of the axis of rotation of a horizontal single axis tracker can be shared between trackers to lower the installation cost. This type of solar tracker is most appropriate for low latitude regions. Field layouts with horizontal single axis trackers are very flexible. The simple geometry means that keeping all of the axes of rotation parallel to one another is all that is required for appropriately positioning the trackers with respect to one another. Appropriate spacing can maximize the ratio of energy production to cost, this being dependent upon local terrain and shading conditions and the time-of-day value of the energy produced. Backtracking is one means of computing the disposition of panels. Horizontal trackers typically have the face of the module oriented parallel to the axis of rotation. As a module tracks, it sweeps a cylinder that is rotationally symmetric around the axis of rotation. In single axis horizontal trackers, a long horizontal tube is supported on bearings mounted upon pylons or frames. The axis of the tube is on a north–south line. Panels are mounted upon the tube, and the tube will rotate on its axis to track the apparent motion of the Sun through the day.

Horizontal Single Axis tracker with Tilted Modules
in Xitieshan, China. Commissioned in July 2014.

Horizontal Single Axis Tracker with Tilted Modules (HTSAT)

In HSAT, the modules are mounted flat at 0 degrees, while in HTSAT, the modules are installed at a certain tilt. It works on same principle as HSAT, keeping the axis of tube horizontal in north–south line and rotates the solar modules east to west throughout the day. These trackers are usually suitable in high latitude locations but does not take as much land space as consumed by Vertical single axis tracker (VSAT). Therefore, it brings the advantages of VSAT in a horizontal tracker and minimizes the overall cost of solar project.

Vertical

Vertical Single Axis Tracker (VSAT)

The axis of rotation for vertical single axis trackers is vertical with respect to the ground. These trackers rotate from East to West over the course of the day. Such trackers are more effective at high latitudes than are horizontal axis trackers. Field layouts must consider shading to avoid unnecessary energy losses and to optimize land utilization. Also optimization for dense packing is limited due to the nature of the shading over the course of a year. Vertical single axis trackers typically have the face of the module oriented at an angle with respect to the axis of rotation. As a module tracks, it sweeps a cone that is rotationally symmetric around the axis of rotation.

Tilted

Tilted Single Axis Tracker (TSAT)

All trackers with axes of rotation between horizontal and vertical are considered tilted single axis trackers. Tracker tilt angles are often limited to reduce the wind profile and decrease the elevated end height. With backtracking, they can be packed without shading perpendicular to their axis of rotation at any density. However, the packing parallel to their axes of rotation is limited by the tilt angle and the latitude. Tilted single axis trackers typically have the face of the module oriented parallel to the axis of rotation. As a module tracks, it sweeps a cylinder that is rotationally symmetric around the axis of rotation.

Tilted single axis tracker in Siziwangqi, China.

Dual Axis Trackers

Dual axis trackers have two degrees of freedom that act as axes of rotation. These axes are typically normal to one another. The axis that is fixed with respect to the ground can be considered a primary axis. The axis that is referenced to the primary axis can be considered a secondary axis. There are several common implementations of dual axis trackers. They are classified by the orientation of their primary axes with respect to the ground. Two common implementations are tip-tilt dual axis trackers (TTDAT) and azimuth-altitude dual axis trackers (AADAT). The orientation of the module with respect to the tracker axis is important when modeling performance. Dual axis trackers typically have modules oriented parallel to the secondary axis of rotation. Dual axis trackers allow for optimum solar energy levels due to their ability to follow the Sun vertically and horizontally. No matter where the Sun is in the sky, dual axis trackers are able to angle themselves to be in direct contact with the Sun.

Tip–tilt

Dual axis tracker mounted on a pole. Project in Siziwangqi.

A tip–tilt dual axis tracker (TTDAT) is so-named because the panel array is mounted on the top of a pole. Normally the east–west movement is driven by rotating the array around the top of the pole. On top of the rotating bearing is a T- or H-shaped mechanism that provides vertical rotation of the panels and provides the main mounting points for the array. The posts at either end of the primary axis of rotation of a tip–tilt dual axis tracker can be shared between trackers to lower installation costs.

Other such TTDAT trackers have a horizontal primary axis and a dependent orthogonal axis. The vertical azimuthal axis is fixed. This allows for great flexibility of the payload connection to the ground mounted equipment because there is no twisting of the cabling around the pole.

Field layouts with tip–tilt dual axis trackers are very flexible. The simple geometry means that keeping the axes of rotation parallel to one another is all that is required for appropriately positioning the trackers with respect to one another. Normally the trackers would have to be positioned at fairly low density in order to avoid one tracker casting a shadow on others when the Sun is low in the sky. Tip-tilt trackers can make up for this by tilting closer to horizontal to minimize up-Sun shading and therefore maximize the total power being collected.

The axes of rotation of many tip–tilt dual axis trackers are typically aligned either along a true north meridian or an east–west line of latitude.

Given the unique capabilities of the Tip-Tilt configuration and the appropriate controller, totally automatic tracking is possible for use on portable platforms. The orientation of the tracker is of no importance and can be placed as needed.

Azimuth-altitude dual axis tracker - 2 axis solar tracker, Toledo, Spain.

Azimuth-altitude

An azimuth–altitude (or alt-azimuth) dual axis tracker (AADAT) has its primary axis (the azimuth axis) vertical to the ground. The secondary axis, often called elevation axis, is then typically normal to the primary axis. They are similar to tip-tilt systems in operation, but they differ in the way the array is rotated for daily tracking. Instead of rotating the array around the top of the pole, AADAT systems can use a large ring mounted on the ground with the array mounted on a series of rollers. The main advantage of this arrangement is the weight of the array is distributed over a portion of the ring, as opposed to the single loading point of the pole in the TTDAT. This allows AADAT to support much larger arrays. Unlike the TTDAT, however, the AADAT system cannot be placed closer together than the diameter of the ring, which may reduce the system density, especially considering inter-tracker shading.

Construction and Self-build

The economical balance between cost of panel and tracker is not trivial. The steep drop in cost for solar panels in the early 2010s made it more challenging to find a sensible solution. As can be seen in the attached media files, most constructions use industrial and/or heavy materials unsuitable for small or craft workshops. Even commercial offers may have rather unsuitable solutions (a big rock) for stabilisation. For a small(amateur/enthusiast) construction following criteria have to be met: economy, stability of endproduct against elemental hazards, ease of handling materials and joinery.

Tracker Type Selection

The selection of tracker type is on many factors including installation size, electric rates, government incentives, land constraints, latitude, and local weather.

Horizontal single axis trackers are typically used for large distributed generation projects and utility scale projects. The combination of energy improvement and lower product cost and lower installation complexity results in compelling economics in large deployments. In addition the strong afternoon performance is particularly desirable for large grid-tied photovoltaic systems so that production will match the peak demand time. Horizontal single axis trackers also add a substantial amount of productivity during the spring and summer seasons when the Sun is high in the sky. The inherent robustness of their supporting structure and the simplicity of the mechanism also result in high reliability which keeps maintenance costs low. Since the panels are horizontal, they can be compactly placed on the axle tube without danger of self-shading and are also readily accessible for cleaning.

A vertical axis tracker pivots only about a vertical axle, with the panels either vertical, at a fixed, adjustable, or tracked elevation angle. Such trackers with fixed or (seasonally) adjustable angles are suitable for high latitudes, where the apparent solar path is not especially high, but which leads to long days in summer, with the Sun traveling through a long arc.

Dual axis trackers are typically used in smaller residential installations and locations with very high government feed in tariffs.

Multi-mirror Concentrating PV

Reflective mirror concentrator units.

This device uses multiple mirrors in a horizontal plane to reflect sunlight upward to a high temperature photovoltaic or other system requiring concentrated solar power. Structural problems and expense are greatly reduced since the mirrors are not significantly exposed to wind loads. Through the employment of a patented mechanism, only two drive systems are required for each device. Because of the configuration of the device it is especially suited for use on flat roofs and at lower latitudes. The units illustrated each produce approximately 200 peak DC watts.

A multiple mirror reflective system combined with a central power tower is employed at the Sierra SunTower, located in Lancaster, California. This generation plant operated by eSolar is scheduled to begin operations on August 5, 2009. This system, which uses multiple heliostats in a north–south alignment, uses pre-fabricated parts and construction as a way of decreasing startup and operating costs.

Drive Types

Active Tracker

Active trackers use motors and gear trains to perform solar tracking. They can use microprocessors and sensors, date and time-based algorithms, or a combination of both to detect the position of the sun. In order to control and manage the movement of these massive structures special slewing drives are designed and rigorously tested. The technologies used to direct the tracker are constantly evolving and recent developments at Google and Eternegy have included the use of wire-ropes and winches to replace some of the more costly and more fragile components.

A slewing drive gearbox.

Counter rotating slewing drives sandwiching a fixed angle support can be applied to create a "multi-axis" tracking method which eliminates rotation relative to longitudinal alignment. This method if placed on a column or pillar will generate more electricity than fixed PV and its PV array will never rotate into a parking lot drive lane. It will also allow for maximum solar generation in virtually any parking lot lane/row orientation, including circular or curvilinear.

Active two-axis trackers are also used to orient heliostats - movable mirrors that reflect sunlight toward the absorber of a central power station. As each mirror in a large field will have an individual orientation these are controlled programmatically through a central computer system, which also allows the system to be shut down when necessary.

Light-sensing trackers typically have two or more photosensors, such as photodiodes, configured differentially so that they output a null when receiving the same light flux. Mechanically, they should be omnidirectional (i.e. flat) and are aimed 90 degrees apart. This will cause the steepest part of their cosine transfer functions to balance at the steepest part, which translates into maximum sensitivity.

Since the motors consume energy, one wants to use them only as necessary. So instead of a continuous motion, the heliostat is moved in discrete steps. Also, if the light is below some threshold there would not be enough power generated to warrant reorientation. This is also true when there is not enough difference in light level from one direction to another, such as when clouds are passing overhead. Consideration must be made to keep the tracker from wasting energy during cloudy periods.

Passive Tracker

Passive tracker head in spring/summer tilt position with panels on light blue rack pivoted to morning position against stop; dark blue objects are hydraulic dampers.

The most common Passive trackers use a low boiling point compressed gas fluid that is driven to one side or the other (by solar heat creating gas pressure) to cause the tracker to move in response to an imbalance. As this is a non-precision orientation it is unsuitable for certain types of concentrating photovoltaic collectors but works fine for common PV panel types. These will have viscous dampers to prevent excessive motion in response to wind gusts. Shader/reflectors are used to reflect early morning sunlight to "wake up" the panel and tilt it toward the Sun, which can take some hours dependant on shading conditions. The time to do this can be greatly reduced by adding a self-releasing tiedown that positions the panel slightly past the zenith (so that the fluid does not have to overcome gravity) and using the tiedown in the evening. (A slack-pulling spring will prevent release in windy overnight conditions.)

A newly emerging type of passive tracker for photovoltaic solar panels uses a hologram behind stripes of photovoltaic cells so that sunlight passes through the transparent part of the module and reflects on the hologram. This allows sunlight to hit the cell from behind, thereby increasing the module's efficiency. Also, the panel does not have to move since the hologram always reflects sunlight from the correct angle towards the cells.

Manual Tracking

In some developing nations, drives have been replaced by operators who adjust the trackers. This has the benefits of robustness, having staff available for maintenance and creating employment for the population in the vicinity of the site.

Rotating Buildings

In Freiburg im Breisgau, Germany, Rolf Disch built the Heliotrop in 1996, a residential building that is rotating with the sun and has an additional dual axis photovoltaic sail on the roof. It's producing four times the amount of energy the building consumes.

The Gemini House is a unique example of a vertical axis tracker. This cylindrical house in Austria (latitude above 45 degrees north) rotates in its entirety to track the Sun, with vertical solar panels mounted on one side of the building, rotating independently, allowing control of the natural heating from the Sun.

Gemini house rotates in its entirety.

ReVolt House is a rotating, floating house designed by TU Delft students for the Solar Decathlon Europe competition in Madrid. The house would be realized in September 2012. A closed façade turns itself towards the Sun in summer to prevent the interior space from direct heat gains. In winter, the glass façade faces the Sun to get direct sunlight in the house.

Disadvantages

Trackers add cost and maintenance to the system - if they add 25% to the cost, and improve the output by 25%, the same performance can be obtained by making the system 25% larger, eliminating the additional maintenance. Tracking was very cost effective in the past when photovoltaic modules were expensive compared to today. Because they were expensive, it was important to use tracking to minimize the number of panels used in a system with a given power output. But as panels get cheaper, the cost effectiveness of tracking vs using a greater number of panels decreases. However, in off-grid installations where batteries store power for overnight use, a tracking system reduces the hours that stored energy is used thus requiring less battery capacity. As the batteries themselves are expensive (either traditional lead acid stationary cells or newer lithium ion batteries) then this cost need adding into the cost effectiveness equation.

Tracking is also not suitable for typical residential rooftop photovoltaic installations. Since tracking requires that panels tilt or otherwise move, provisions must be made to allow this. This requires that panels be offset a significant distance from the roof, which requires expensive racking and increases wind load. Also, such a setup would not make for a very aesthetically pleasing install on residential rooftops. Because of this (and the high cost of such a system), tracking is not used on residential rooftop installations, and is unlikely to ever be used in such installations. This is especially true as the cost of photovoltaic modules continues to decrease, which makes increasing the number of modules for more power the more cost-effective option. Tracking can (and sometimes is) used for residential ground mount installations, where greater freedom of movement is possible.

Tracking can also cause shading problems. As the panels move during the course of the day, it is possible that, if the panels are located too close to one another, they may shade one another due to profile angle effects. As an example, if you have several panels in a row from east to west, there will be no shading during solar noon. But in the afternoon, panels could be shaded by their west neighboring panel if they are sufficiently close. This means that panels must be spaced sufficiently far to prevent shading in systems with tracking, which can reduce the available power from a given area during the peak Sun hours. This is not a big problem if there is sufficient land area to widely space the panels. But it will reduce output during certain hours of the day (i.e. around solar noon) compared to a fixed array.

Further, single-axis tracking systems are prone to become unstable already at relatively modest wind speeds (galloping). This is due to torsional instability of single-axis solar tracking systems. Anti-galloping measures such as automatic stowing and external dampers must be implemented.

SOLAR INVERTER

The Solar Inverter is an essential device in any solar power system. Its basic function of the inverter is to change the variable Direct Current output of the solar panels into Alternating Current. The various electrical and electronic components connected in the circuit help in the conversion.

Solar Inverter.

The converted Alternating Current power is used for running your appliances like the TV, Refrigerator, Microwave, etc. For some particular applications, we can directly use the Direct Current power from the solar panel such as LED night lights, a cell phone charger. Generally, the power of a home solar power system is used for power AC loads.

Working Principle of a Solar Inverter

The inverters work by taking in power from a Direct Current (DC) Source, i.e., the solar panel. The power is generated in the range of 250 Volts to 600 Volts. DC power is converted into AC power by the inversion process taking place in the inverter. This process of DC to AC Conversion is achieved by using a set of solid state devices like Insulated Gate Bipolar Transistors (IGBT's.). These devices when connected in a typical H-Bridge arrangement oscillate the DC power thereby creating AC power.

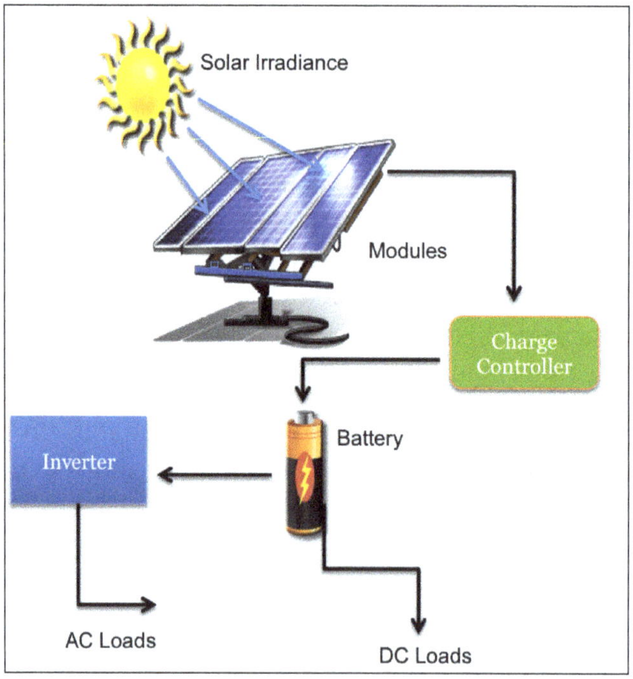

Working of an Inverter.

A step up transformer is used in between the devices. At this stage, the AC power output is obtained, which is either consumed by the local loads or can be fed to the grid. A few manufacturers have started manufacturing transformer-less inverters which have higher efficiencies as compared to inverters having transformers.

In any solar power inverter a micro-controller chip is present which is programmed to execute various complex algorithms in a precise manner. The controller maximizes the power output from the photovoltaic panels through the Maximum Power Point Tracking (MPPT) algorithm.

Types of Solar Inverters

There are abundant of well-known creators of Solar Inverters all over the world. But there are a few types of solar inverters available in the market which includes the following:

Off Grid Inverters

Off grid inverters are used in remote systems wherein the solar inverter is fed DC power from a battery panel.This battery panel is charged by solar panels.Several such inverters have integrated with basic battery chargers which can be used to boost the battery from an AC power source.

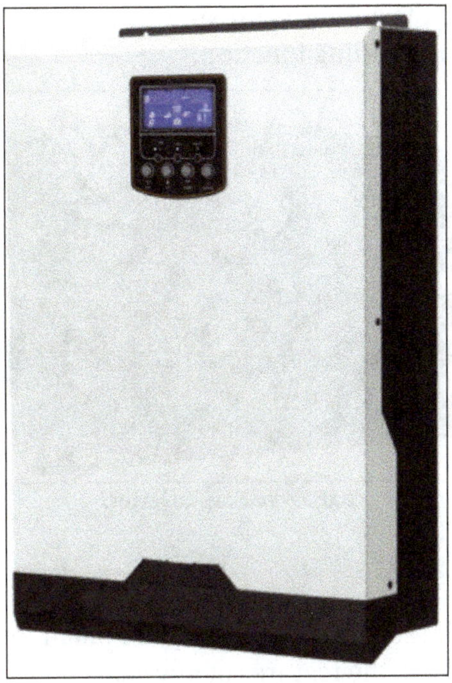

Off Grid Inverter.

Grid Tie Inverters

An inverter which is associated with the grid can be said to be a grid tie inverter. These inverter feed power in the electricity grid by corresponding phase and frequency. And, the frequency ranges 50Hz in India and 60Hz in North America of the o/p AC power with the effectiveness supplied AC power. These inverters are designed to automatically shut down once sensing a loss of supply from the utility.

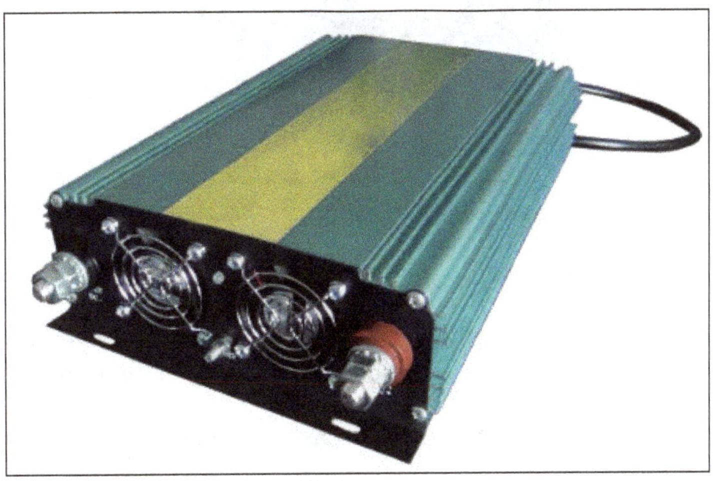

Grid Tie Inverter.

Battery Backup Inverters

These special types of inverters are particularly designed to draw energy from a battery. The charge of the battery is preserved by using an on-board charger and an additional energy is transferred to the grid. Such inverters have the ability to provide AC power to particular loads during power outages. They also have the anti-islanding function.

Battery backup Inverter.

Micro Inverters

Microinverters are modern in the solar industry. They are small, compact size and portable with a lot of performance. They include all the features of any central inverters.

Micro Inverter.

Advantages of Solar Inverter

After knowing in detail what is a solar inverter and how it is a suitable making appliance work at residential & industrial levels we must discuss about the several advantages of the solar inverter:

- Solar energy has constantly helped in decreasing the greenhouse effect and global warming.
- By using of solar based devices will help in saving money and also energy. Because many people have started using these devices.
- A solar inverter helps in changing the DC into batteries or AC. This supports people who use a partial amount of electricity.

- The synchronous solar inverter that helps small homeowners and also power companies as they are huge in size.
- The multifunction solar inverter is the finest among all and works powerfully. It converts the DC to AC very carefully which is suitable for commercial establishments.
- This inverter is cost effective, i.e. low cost than generators.
- Apart from these there are additional devices too, that make use of solar energy such as, solar heater, cooker.

Disadvantages of Solar Inverter

- Primarily, we need to shell out a lot of money for purchasing a solar inverter.
- It will work efficiently and produce DC only when the daylight is strong.
- The solar panels are used to attract the sunshine needs lots of space.
- Solar Inverters can work when there is no Sunshine but the battery which is available in that is charged fully with the help of Sunshine.

References

- B Green, M. A. (2003), "Crystalline and thin-film silicon solar cells: state of the art and future potential", Solar Energy, 74 (3): 181–192, Bibcode:2003soen...74..181G, doi:10.1016/S0038-092X(03)00187-7
- Solar-cell: newworldencyclopedia.org, Retrieved 13 May, 2019
- Spray-deposition steers perovskite solar cells towards commercialisation". Chemistryworld. July 29, 2014. Retrieved August 4, 2015
- Delamarre; et al. (2014). "Quantitative luminescence mapping of Cu(In,Ga)Se2 thin-film solar cells". Progress in Photovoltaics. 23 (10): 1305–1312. Doi:10.1002/pip.2555
- New South Innovations News - UNSW breaks solar cell record". Newsouth Innovations. 2008-11-18. Archived from the original on April 25, 2012. Retrieved 2012-06-23
- Lukas, Schmidt-Mende (2016-05-24). Organic and hybrid solar cells : an introduction. Weickert, Jonas. Berlin. ISBN 9783110283204. OCLC 950902053
- What-is-a-solar-panel: mrsolar.com, Retrieved 25 February, 2019
- Brown, A; Green, M (2002). "Detailed balance limit for the series constrained two terminal tandem solar cell". Physica E. 14 (1–2): 96–100. Bibcode:2002phye...14...96B. Doi:10.1016/S1386-9477(02)00364-8
- How-solar-inverter-works-applications: watelectrical.com, Retrieved 16 January, 2019

Electricity Generation Systems

CHAPTER 4

The process that generates electric power from primary energy sources is referred to as electricity generation. Some of the electricity generation systems are photovoltaic power stations, grid-connected photovoltaic power systems, rooftop photovoltaic power station and building-integrated photovoltaics. This chapter closely examines these types of electricity generation systems to provide an extensive understanding of the subject.

PHOTOVOLTAIC POWER STATION

A photovoltaic power plant or a photovoltaic power station is a set of facilities designed to supply electricity to the grid through the use of large-scale photovoltaic systems. The function of the photovoltaic power station is to capture and transform solar radiation into electricity.

A photovoltaic power plant is a power plant consisting of photovoltaic modules and an inverter. The photovoltaic panels are in charge of transforming the solar radiation, into electrical energy of direct current. The inverter is the electronic equipment whose function is to convert the direct current power produced by the photovoltaic modules into alternating current power with similar characteristics to those of the electric grid.

The solar installation works as a small electricity production plant, which, through the inverter, injects the energy produced into the electricity distribution network.

This operation entails a very important advantages: a better performance of the installation. The energy produced is not used directly but is injected into the network and can be used by any subscriber of the electrical system. In this way, this allows all the energy generated by the solar modules to be used, even on holidays or holiday periods where it can be practically nil.

Components of a Photovoltaic Power Plant

A photovoltaic power plant is composed of different components, of which we mention the main ones:

- Photovoltaic cells. Generally composed of silicon. The photoelectric cells are responsible for capturing solar energy and transforming it into electricity through the photovoltaic effect.
- Weather tower. It is the place where the different meteorological conditions are analyzed to determine the solar radiation that is being received or expected to be received.
- Direct current cabinet. Receive the electricity generated by the photovoltaic cells.
- Investor. Convert the direct current to alternating current.
- Alternating current cabinet. It receives the electricity that the investor has transformed into alternating current.
- Transformation center. Site where the energy adapts to the conditions of intensity and voltage apt to be transported.
- Transportation lines. These are the lines that allow electric energy to be transported to consumption centers.
- Control room. Site where the operation of all the elements of the photovoltaic power plant is supervised.

Operation of a Photovoltaic Power Station

The operation of all the equipment of the plant is supervised from the control room. In the control room, information is received from the different systems of the installation: meteorological tower, inverter, direct current DC and alternating current AC cabinets, transformation center, etc.

Generation of Electricity through Photovoltaic Panels

The set of photovoltaic cells, grouped in photovoltaic panels, is the basic element of a photovoltaic power plant. The photeléctricas cells are responsible for capturing the Sun's radiation and transforming it into electrical energy. These photovoltaic cells are usually composed of silicon, which is a semiconductor material that facilitates the photoelectric effect. These cells are responsible for generating a continuous current of electrons that will later be transformed into alternating current.

The photovoltaic cells are integrated in modules and then, with these modules, the photovoltaic panels are formed.

A fundamental element for a photovoltaic power station is meteorology (radiation, humidity, temperature). Depending on the meteorological situation of each moment, the solar radiation that the photovoltaic cells will receive will be variable. For this reason a meteorological tower is built in the plant to control and analyze the different weather conditions.

Transformation of Direct Current to Alternating Current

The photovoltaic panels generate direct current, but as the electric power that circulates through the transport network it does so in the form of alternating current. For this reason, the direct current must be converted to alternating current.

First, the direct current from the solar panels is led to a direct current cabinet. In this cabinet the current is transformed into alternating current by means of an inverter. Next, the current is transported to an alternating current cabinet.

Transport and Electricity Supply from the Photovoltaic Power Plant

The current that arrives in the alternating current cabinet is not able to be supplied to the network. For this reason, the electricity produced passes through a transformation center where it adapts to the intensity and tension conditions of the transport lines for use in the consumption centers.

GRID-CONNECTED PHOTOVOLTAIC POWER SYSTEM

A grid-connected, residential solar rooftop system near Boston, USA.

A grid-connected photovoltaic system, or grid-connected PV system is an electricity generating solar PV power system that is connected to the utility grid. A grid-connected PV system consists of solar panels, one or several inverters, a power conditioning unit and grid connection equipment. They range from small residential and commercial rooftop systems to large utility-scale solar power stations. Unlike stand-alone power systems, a grid-connected system rarely includes an integrated battery solution, as they are still very expensive. When conditions are right, the grid-connected PV system supplies the excess power, beyond consumption by the connected load, to the utility grid.

Operation

Photovoltaic power station at Nellis Air Force Base, United States.

Residential, grid-connected rooftop systems which have a capacity more than 10 kilowatts can meet the load of most consumers. They can feed excess power to the grid where it is consumed by other users. The feedback is done through a meter to monitor power transferred. Photovoltaic wattage may be less than average consumption, in which case the consumer will continue to purchase grid energy, but a lesser amount than previously. If photovoltaic wattage substantially exceeds average consumption, the energy produced by the panels will be much in excess of the demand. In this case, the excess power can yield revenue by selling it to the grid. Depending on their agreement with their local grid energy company, the consumer only needs to pay the cost of electricity consumed less the value of electricity generated. This will be a negative number if more electricity is generated than consumed. Additionally, in some cases, cash incentives are paid from the grid operator to the consumer.

Connection of the photovoltaic power system can be done only through an interconnection agreement between the consumer and the utility company. The agreement details the various safety standards to be followed during the connection.

Features

Electric power from photovoltaic panels must be converted to alternating current by a power inverter if it is intended for delivery to a power grid. The inverter sits between the solar array and the grid, and may be a large stand-alone unit or may be a collection of small inverters attached to individual solar panels as an AC module. The inverter must monitor grid voltage, waveform, and frequency. The inverter must detect failure of the grid supply and must not supply power to the grid.

An inverter connected to a malfunctioning power line will automatically disconnect in accordance with safety rules, which vary by jurisdiction. The location of the fault current plays a crucial part in deciding whether the protection mechanism of the inverter will kick in, especially for low and medium electricity supply network. A protection system must ensure proper operation for faults external to the inverter on the supply network. The inverter must be designed to synchoronize its AC frequency with the grid, to ensure correct direction of power flow.

Anti-islanding

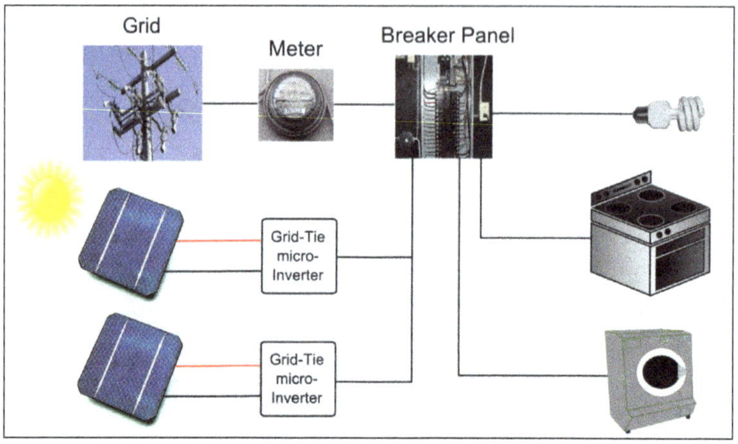

Diagram of a residential grid-connected PV system.

Islanding is the condition in which a distributed generator continues to power a location even though power from the electric utility grid is no longer present. Islanding can be dangerous to utility workers, who may not realize that a circuit is still powered, even though there is no power from the electrical grid. For that reason, distributed generators must detect islanding and immediately stop producing power; this is referred to as anti-islanding.

In the case of a utility blackout in a grid-connected PV system, the solar panels will continue to deliver power as long as the sun is shining. In this case, the supply line becomes an "island" with power surrounded by a "sea" of unpowered lines. For this reason, solar inverters that are designed to supply power to the grid are generally required to have automatic anti-islanding circuitry in them. In intentional islanding, the generator disconnects from the grid, and forces the distributed generator to power the local circuit. This is often used as a power backup system for buildings that normally sell their power to the grid.

There are two types of anti-islanding control techniques:

- Passive: The voltage and/or the frequency change during the grid failure is measured and a positive feedback loop is employed to push the voltage and/or the frequency further away from its nominal value. Frequency or voltage may not change if the load matches very well with the inverter output or the load has a very high quality factor (reactive to real power ratio). So there exists some Non Detection Zone (NDZ).
- Active: This method employs injecting some error in frequency or voltage. When grid fails, the error accumulates and pushes the voltage and/or frequency beyond the acceptable range.

Advantages

- Systems such as Net Metering and Feed-in Tariff which are offered by some system operators, can offset a customers electricity usage costs. In some locations though, grid technologies cannot cope with distributed generation feeding into the grid, so the export of surplus electricity is not possible and that surplus is earthed.
- Grid-connected PV systems are comparatively easier to install as they do not require a battery system.
- Grid interconnection of photovoltaic (PV) power generation systems has the advantage of effective utilization of generated power because there are no storage losses involved.
- A photovoltaic power system is carbon negative over its lifespan, as any energy produced over and above that to build the panel initially offsets the need for burning fossil fuels. Even though the sun doesn't always shine, any installation gives a reasonably predictable average reduction in carbon consumption.

Disadvantages

- Grid-connected PV can cause issues with voltage regulation. The traditional grid operates under the assumption of one-way, or radial, flow. But electricity injected into the grid increases voltage, and can drive levels outside the acceptable bandwidth of ±5%.
- Grid-connected PV can compromise power quality. PV's intermittent nature means rapid changes in voltage. This not only wears out voltage regulators due to frequent adjusting, but also can result in voltage flicker.
- Connecting to the grid poses many protection-related challenges. In addition to islanding, too high levels of grid-connected PV result in problems like relay desensitization, nuisance tripping, interference with automatic reclosers, and ferroresonance.

SOLAR ARK

The Solar Ark is an ark-shaped solar photovoltaic power generation facility which offers activities to cultivate a better appreciation of solar power generation, and thereby benefitting both ecology and science. This 315-meter-wide, 37-meter-tall facility is located in Anpachi, Gifu Prefecture, in the geographical center of Japan, and can be seen from the JR Tōkaidō bullet train, which runs past on an adjacent railway. It has over 5000 panels that produce approximately 530,000 kilowatt-hours on an annual basis and a maximum system power of 630 kilowatts.

Stationed at the center of the Solar Ark is the Solar Lab, a museum of solar energy. A hands-on, outdoor light exhibition was planned for opening in 2005. The Solar Ark was an enterprise partner with the 2005 World Exhibition, Aichi Prefecture, Japan. It is one of the largest solar buildings in the world.

Design

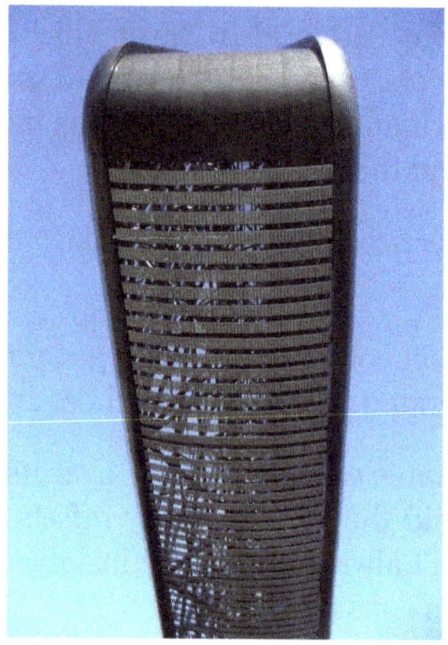

The Solar Ark viewed from below.

The Solar Ark's design was inspired by the vision of an ark embarking on a journey to the 21st century. This idea led to the Solar Ark's size and overall symbolic shape of being an example of producing clean energy. In total, the construction area for the Solar Ark is 3294.48 m^2 reinforced concrete was used for the base of the construction. From one end to the other, the total length of the Solar Ark is 315 metres. The ark is 31.6 metres tall from the centre of the structure. There are 5,046 solar panels in total. Twelve single-crystal silicon solar cell modules per unit were assembled on the ground, and 470 units were lifted up and attached to the main body of the Solar Ark.

Panasonic Solar Ark.

The weight of the actual body of the ark (pillars being excluded) is 3,000 metric tonnes and is constructed from structural steel. This construction material helps give the impression of the solar ark being suspended in the air. Each column is 2 metres in diameter and 31 meters in length and the Solar Ark is 315 metres long. The entire Solar Ark chassis is supported by four "G-Columns" which are custom built pillars by Kubota and in total, these pillars weigh approximately 5,000 tonnes. These high-quality pillars are homogeneous, the result of the seamless method of construction that utilizes centrifugal force. Due to the Solar Ark's sturdy building materials, the Ark is able to resist winds of up to 34 meters/second and level 7 earthquakes on the Japanese scale.

The ark is surrounded by 5-meter high water fountains and two ponds, each having their own cascade. The entrance to the Solar Ark has solar wings which are composed of HIT solar cells that generate electricity on the topside and underside while also functioning as awnings that allow sunlight to filter through. Between the individual solar panels, there are, in total, 75,000 red, green, and blue computer controlled LEDs which are activated at night to produce images and words across the ark.

ROOFTOP PHOTOVOLTAIC POWER STATION

Rooftop PV systems around the world: Berlin, Germany.

A rooftop photovoltaic power station, or rooftop PV system, is a photovoltaic system that has its electricity-generating solar panels mounted on the rooftop of a residential or commercial building or structure. The various components of such a system include photovoltaic modules, mounting systems, cables, solar inverters and other electrical accessories.

Bensheim, Germany.

Kuppam, India.

Rooftop mounted systems are small compared to ground-mounted photovoltaic power stations with capacities in the megawatt range. Rooftop PV systems on residential buildings typically feature a capacity of about 5 to 20 kilowatts (kW), while those mounted on commercial buildings often reach 100 kilowatts or more.

Installation

Workers install residential rooftop solar panels.

The urban environment provides a large amount of empty rooftop spaces and can inherently avoid the potential land use and environmental concerns. Estimating rooftop solar insolation is a multi-faceted process, as insolation values in rooftops are impacted by the following:

- Time of the year,
- Latitude,
- Weather conditions,
- Roof slope,
- Roof aspect,
- Shading from adjacent buildings and vegetation.

Rooftop PV systems at Googleplex, California.

There are various methods for calculating potential solar PV roof systems including the use of Lidar and orthophotos. Sophisticated models can even determine shading losses over large areas for PV deployment at the municipal level.

Components of a Rooftop Solar Array

The following part contains the most commonly utilized components of a rooftop solar array. Though designs may vary with roof type (eg. metal vs shingle), roof angle, and shading concerns, most arrays consist of some variation of the following components:

- Solar Panels produce carbon free electricity when irradiated with (sun)light. Often made of Silicon, solar panels are made of smaller solar cells which typically number 6 cells per panel. Multiple solar panels strung together make up a solar array. Solar panels are generally protected by tempered glass and secured with an aluminum frame. The front of a solar panel is very durable whereas the back of a panel is generally more vulnerable.
- Mounting clamps generally consist of aluminum brackets and stainless steel bolts that secure solar panels to one another on the roof and onto the rails. Clamps often vary in design in order to account for various roof and rail configurations.
- Racking or rails are made of metal and often lie in a parallel configuration on the roof for the panels to lie on. It is important that the rails are level enough for the panels to be evenly mounted.
- Mounts attach the rails and the entire array to the surface of the roof. These mounts are often L brackets that are bolted through flashing and into the rafters of the roof. Mounts vary in design due to the wide range of roof configurations and materials.
- Flashings are a durable metal plate that provide a water resistant seal between the mounts and roof surface. Oftentimes, caulk is used to seal the flashing to the roof and it resembles a metal roof shingle.
- DC/AC wiring for inverters connect wires between panels and into a micro inverter or string inverter. No cables should touch the roof surface or hang from the array to avoid weathering and the deterioration of cables.
- Micro inverters are mounted to the bottom of the panel and convert DC power from the panels into AC power that can be sent into the grid. Micro inverters allow for the optimization of each panel when shading occurs and specific data from individual panels.

Feed-in Tariff Mechanism

In a grid connected rooftop photovoltaic power station, the generated electricity can sometimes be sold to the servicing electric utility for use elsewhere in the grid. This arrangement provides payback for the investment of the installer. Many consumers from across the world are switching to this mechanism owing to the revenue yielded. A public utility commission usually sets the rate that the utility pays for this electricity, which could be at the retail rate or the lower wholesale rate, greatly affecting solar power payback and installation demand.

The FIT as it is commonly known has led to an expansion in the solar PV industry worldwide. Thousands of jobs have been created through this form of subsidy. However it can produce a bubble effect which can burst when the FIT is removed. It has also increased the ability for localised production and embedded generation reducing transmission losses through power lines.

Hybrid Systems

Rooftop PV hybrid system.

A rooftop photovoltaic power station (either on-grid or off-grid) can be used in conjunction with other power components like diesel generators, wind turbines, batteries etc. These solar hybrid power systems may be capable of providing a continuous source of power.

Advantages

Installers have the right to feed solar electricity into the public grid and hence receive a reasonable premium tariff per generated kWh reflecting the benefits of solar electricity to compensate for the current extra costs of PV electricity.

Disadvantages

An electrical power system containing a 10% contribution from PV stations would require a 2.5% increase in load frequency control (LFC) capacity over a conventional system - in issue which may be countered by using synchronverters in the DC/AC-circuit of the PV system. The break-even cost for PV power generation was in 1996 found to be relatively high for contribution levels of less than 10%. Higher proportions of PV power generation gives lower break-even costs, but economic and LFC considerations imposed an upper limit of about 10% on PV contributions to the overall power systems.

Technical Challenges

There are many technical challenges to integrating large amounts of rooftop PV systems to the power grid. For example:

- Reverse Power Flow

  The electric power grid was not designed for two way power flow at the distribution level. Distribution feeders are usually designed as a radial system for one way power flow transmitted over long distances from large centralized generators to customer loads at the end of the distribution feeder. Now with localized and distributed solar PV generation on rooftops, reverse flow causes power to flow to the substation and transformer, causing significant challenges. This has adverse effects on protection coordination and voltage regulators.

- Ramp Rates

  Rapid fluctuations of generation from PV systems due to intermittent clouds cause undesirable levels of voltage variability in the distribution feeder. At high penetration of rooftop PV, this voltage variability reduces the stability of the grid due to transient imbalance in load and generation and causes voltage and frequency to exceed set limits if not countered by power controls. That is, the centralized generators cannot ramp fast enough to match the variability of the PV systems causing frequency mismatch in the nearby system. This could lead to blackouts. This is an example of how a simple localized rooftop PV system can affect the larger power grid. The issue is partially mitigated by distributing solar panels over a wide area, and by adding storage.

BUILDING-INTEGRATED PHOTOVOLTAICS

Building-integrated photovoltaics are photovoltaic materials or components that are used in place of traditional building components or materials, especially in building features such as facades, roofs or skylights, and provide solar power for the building.

Considered as ancillary and, at times, a primary electrical power source, they provide other advantages over conventional building materials. The initial cost involved in building integrated photovoltaics can be offset by the reduction in the labor charges involved and the quantity of building materials involved; furthermore, they are more eco-friendly.

One of the fundamental aspects of building-integrated photovoltaics is the reduction in incremental cost of photovoltaic materials or components and the improved life-cycle cost. The components of a building-integrated photovoltaic system include:

- Photovoltaic module.
- Backup power supply system.
- Charge controller.

- Power storage system.
- Support for wiring, hardware and safety disconnects.

Building-integrated photovoltaics can be used for different functions, such as:

- Replacing conventional materials: One of the key benefits of using building-integrated photovoltaic components is the replacement of traditional building materials like roof membrane, facade cladding or skylight glazing, etc. Not only are they lower in cost but also have less environmental impact.
- Power generation: Building-integrated photovoltaic components can generate sufficient power efficiently and effectively for a building. There are different technologies and projects that work in conjunction with these to ensure a rich energy harvest.
- Creation of shade: When used in buildings, they can protect components from unwanted solar heat and thus provide shading benefits.
- Architectural integration: They can provide comparatively more aesthetic appeal than conventional materials. This helps in adding architectural interest to the building.

There are many challenges in using building-integrated photovoltaics, such as limited access, buildup of heat and non-optimal shading and array orientation.

Applications for Building-Integrated Photovoltaics

- Façade: PV can be integrated into the sides of buildings, replacing traditional glass windows with semi-transparent thin-film or crystalline solar panels. These surfaces have less access to direct sunlight than rooftop systems, but typically offer a larger available area. In retrofit applications, PV panels can also be used to camouflage unattractive or degraded building exteriors.
- Rooftops: In these applications, PV material replaces roofing material or, in some cases, the roof itself. Some companies offer an integrated, single-piece solar rooftop made with laminated glass; others offer solar "shingles" which can be mounted in place of regular roof shingles.
- Glazing: Ultra-thin solar cells may be used to create semi-transparent surfaces, which allow daylight to penetrate while simultaneously generating electricity. These are often used to create PV skylights or greenhouses.

Building Design Considerations

A critical part of maximizing the value of a BIPV system is planning for both environmental and structural factors, both of which influence the economics, aesthetics and overall functionality of any solar system.

Environmental Factors

- Insolation: This refers to the average amount of solar radiation received, usually calculated in $kWh/m^2/day$. It is the most common way to describe the amount of solar resources in a particular area.

- Climate & weather conditions: High ambient temperatures can decrease the output of solar systems, and clouds and rainfall patterns can affect system output and maintenance requirements. High levels of air pollution can require regular cleaning to increase efficiency.
- Shading: Trees, nearby buildings and other structures can block the sun, reducing PV system output.
- Latitude: Distance from the equator affects the optimal tilt angle for solar panels to receive solar radiation.

Structural Factors

- Building energy requirements: The design of a BIPV system should take into account whether the building should be able to operate wholly independent of the electrical grid, which requires batteries or other on-site energy storage systems.
- Solar system design: The design of the PV system itself is determined by the buildings energy requirements, and any structural or aesthetic limitations that may limit material choices. Crystalline silicon panels have higher electricity outputs per square meter, but greater costs and design constraints. Thin-film materials generate less electricity per square meter, but are less expensive and may be integrated more easily onto more surfaces.

Building Integrated Photovoltaics (BIPV) System

A complete BIPV system includes:

- The PV modules (which might be thin-film or crystalline, transparent, semi-transparent, or opaque);
- A charge controller, to regulate the power into and out of the battery storage bank (in stand-alone systems);
- A power storage system, generally comprised of the utility grid in utility-interactive systems or, a number of batteries in stand-alone systems;
- Power conversion equipment including an inverter to convert the PV modules' DC output to AC compatible with the utility grid;
- Backup power supplies such as diesel generators (optional-typically employed in stand-alone systems); and
- Appropriate support and mounting hardware, wiring, and safety disconnects.

BIPV systems can either be interfaced with the available utility grid or they may be designed as stand-alone, off-grid systems. The benefits of power production at the point of use include savings to the utility in the losses associated with transmission and distribution (known as 'grid support'), and savings to the consumer through lower electric bills because of peak shaving (matching peak production with periods of peak demand). Moreover, buildings that produce power using renewable energy sources reduce the demands on traditional utility generators, often reducing the overall emissions of climate-change gasses.

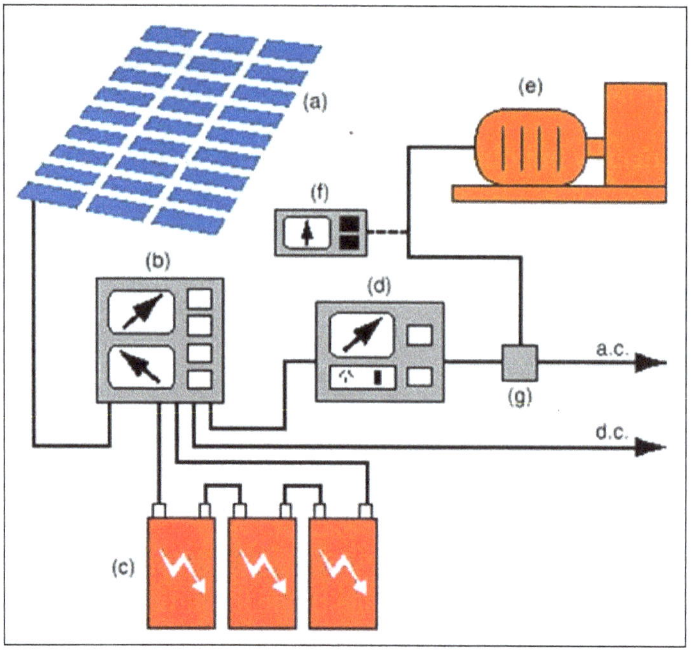

BIPV system diagram.

Design of a Building Integrated Photovoltaics (BIPV) System

BIPV systems should be approached to where energy conscious design techniques have been employed, and equipment and systems have been carefully selected and specified. They should be viewed in terms of life-cycle cost, and not just initial, first-cost because the overall cost may be reduced by the avoided costs of the building materials and labor they replace. Design considerations for BIPV systems must include the building's use and electrical loads, its location and orientation, the appropriate building and safety codes, and the relevant utility issues and costs.

Steps in designing a BIPV system include:

- Carefully consider the application of energy-conscious design practices and/or energy-efficiency measures to reduce the energy requirements of the building. This will enhance comfort and save money while also enabling a given BIPV system to provide a greater percentage contribution to the load.
- Choose between a Utility-Interactive PV System and a Stand-alone PV System:
    - The vast majority of BIPV systems will be tied to a utility grid, using the grid as storage and backup. The systems should be sized to meet the goals of the owner—typically defined by budget or space constraints; and, the inverter must be chosen with an understanding of the requirements of the utility.
    - For those 'stand-alone' systems powered by PV alone, the system, including storage, must be sized to meet the peak demand/lowest power production projections of the building. To avoid over sizing the PV/battery system for unusual or occasional peak loads, a backup generator is often used. This kind of system is sometimes referred to as a "PV-genset hybrid."

- Shift the Peak: If the peak building loads do not match the peak power output of the PV array, it may be economically appropriate to incorporate batteries into certain grid-tied systems to offset the most expensive power demand periods. This system could also act as an uninterruptible power system (UPS).

- Provide adequate ventilation: PV conversion efficiencies are reduced by elevated operating temperatures. This is truer with crystalline silicon PV cells than amorphous silicon thin-films. To improve conversion efficiency, allow appropriate ventilation behind the modules to dissipate heat.

- Evaluate using hybrid PV-solar thermal systems: As an option to optimize system efficiency, a designer may choose to capture and utilize the solar thermal resource developed through the heating of the modules. This can be attractive in cold climates for the pre-heating of incoming ventilation make-up air.

- Consider integrating daylighting and photovoltaic collection: Using semi-transparent thin-film modules, or crystalline modules with custom-spaced cells between two layers of glass, designers may use PV to create unique daylighting features in facade, roofing, or skylight PV systems. The BIPV elements can also help to reduce unwanted cooling load and glare associated with large expanses of architectural glazing.

- Incorporate PV modules into shading devices: PV arrays conceived as "eyebrows" or awnings over view glass areas of a building can provide appropriate passive solar shading. When sunshades are considered as part of an integrated design approach, chiller capacity can often be smaller and perimeter cooling distribution reduced or even eliminated.

- Design for the local climate and environment: Designers should understand the impacts of the climate and environment on the array output. Cold, clear days will increase power production, while hot, overcast days will reduce array output;/p>.

    - Surfaces reflecting light onto the array (e.g., snow) will increase the array output;

    - Arrays must be designed for potential snow- and wind-loading conditions;

    - Properly angled arrays will shed snow loads relatively quickly;

    - Arrays in dry, dusty environments or environments with heavy industrial or traffic (auto, airline) pollution will require washing to limit efficiency losses.

- Address site planning and orientation issues: Early in the design phase, ensure that your solar array will receive maximum exposure to the sun and will not be shaded by site obstructions such as nearby buildings or trees. It is particularly important that the system be completely unshaded during the peak solar collection period consisting of three hours on either side of solar noon. The impact of shading on a PV array has a much greater influence on the electrical harvest than the footprint of the shadow.

- Consider array orientation: Different array orientation can have a significant impact on the annual energy output of a system, with tilted arrays generating 50%–70% more electricity than a vertical facade.

- Reduce building envelope and other on-site loads: Minimize the loads experienced by the BIPV system. Employ daylighting, energy-efficient motors, and other peak reduction strategies whenever possible.

In addition, BIPV systems can be designed to blend with traditional building materials and designs, or they may be used to create a high-technology, future-oriented appearance. Semi-transparent arrays of spaced crystalline cells can provide diffuse, interior natural lighting. High profile systems can also signal a desire on the part of the owner to provide an environmentally conscious work environment.

References

- Large-Scale Solar Power Generation Facility 'Solar Ark'". Panasonic Corporation of North America. Archived from the original on May 18, 2013. Retrieved January 11, 2013
- Building-integrated-photovoltaics, initiatives: seia.org, Retrieved 29 June, 2019
- Photovoltaic-plant, definitions: solar-energy.technology, Retrieved 14 July, 2019
- "International Guideline For The Certification Of Photovoltaic System Components and Grid-Connected Systems". Iea-pvps.org. Retrieved 2011-07-21
- Building-integrated-photovoltaics, resources: wbdg.org, Retrieved 11 January, 2019
- Nguyen, Ha T.; Pearce, Joshua M. (2012). "Incorporating shading losses in solar photovoltaic potential assessment at the municipal scale". Solar Energy. 86 (5): 1245–1260. Doi:10.1016/j.solener.2012.01.017
- Energy Resources and Resource Criteria". Greenip.org. Archived from the original on 2013-08-28. Retrieved 2011-06-20
- Building-integrated-photovoltaics, definition: techopedia.com, Retrieved 18 April, 2019

Applications

CHAPTER 5

Photovoltaics are widely used in solar-powered refrigerators, solar lamps, solar-powered fans, solar vehicles, solar panels on spacecrafts, solar-powered pumps, solar-powered watches, solar-powered flashlights and solar air conditioners. All these diverse applications of photovoltaics have been carefully analyzed in this chapter.

The applications of photovoltaic solar energy are many and varied. In this field, they include from large power generation plants through photovoltaic panels, to small solar calculators.

A first way to classify the applications of photovoltaic solar energy is to distinguish the applications connected to the electrical network and the isolated installations.

The use of photovoltaic panels in isolated buildings is very useful since the investment needed to place solar panels on the roof of a farm, a chalet in the mountain or in a hotel in a secluded spot, is much less than what it would cost take the electric network there. In this case, it has the advantage of being a renewable energy and, therefore, does not require fuel supply.

Photovoltaic Applications in Systems Connected to the Electrical Network

The photovoltaic installations connected to the network are one of the latest and most innovative applications of photovoltaic systems. This application consists of the installation of photovoltaic panels and an inverter capable of transforming the energy supplied by photovoltaic solar panels and injecting it into the electricity grid.

The investor in this type of installation is the central core. The investor must have certain protections in situations that may occur in the electricity network (for example, network lag, network cut, voltage out of range).

The cost per kWh generated is usually greater than the energy generated by other energy generation systems. For this reason, in order to achieve the economic viability of these systems, the country must have legal regulations and financial aid lines, which offset the higher cost per Kwh generated.

Within the solar energy installations connected to the network, we include large solar installations. These are large areas of land with a large number of photovoltaic panels with the aim of generating large amounts of electricity.

The photovoltaic panels generate electricity in direct current from solar radiation. However, in solar installations connected to the network, this electrical energy must be passed to alternating current.

Photovoltaic Applications in Isolated Systems

In isolated systems, that is, in systems that are not connected to the electricity grid, the diversity of applications of photovoltaic energy is almost infinite. Even so, we highlight a few examples:

Photovoltaic Systems for Lighting

Public lighting, by means of photovoltaic systems, is presented as one of the most economical solutions, to illuminate the entrances in towns, crossroads, rest areas, etc.

This type of photovoltaic applications can also be found in billboards, bus stops, street lights and tunnel lighting among other examples.

Signaling by Photovoltaic Energy

Photovoltaic solar energy allows the automation of headlights. For aerial use, panels are being used to supply beacons and signposts on the slopes. Lighthouses and buoys for maritime use.

Another great application is the signaling of roundabouts, curves, traffic signs, obstacles, etc. by means of high luminosity LEDs. The low consumption of the LEDs allows to realize a small-sized photovoltaic installation in these systems.

In this sense, we also find examples of hour and temperature indicators on public roads, at railway level crossings or on oil platforms.

Another great application, which has recently been added with great importance in road safety, is the signaling of roundabouts, curves, traffic signs, obstacles, etc. by means of high brightness LEDs, which allows a small-sized photovoltaic installation due to its low consumption.

Photovoltaic Systems Related to Telecommunications

In the field of telecommunications, photovoltaic solar energy is used in the fields of mobile telephony, in radio and television repeaters. On the road, the SOS poles are powered by this type of renewable energy since it saves the work of getting them network power cables.

Radars, radiotelephony in general and for military or forest monitoring posts, rural telephony via satellite, are other examples in which solar energy is used to operate.

Rural Photovoltaic Electrification

Rural electrification currently has all the comforts that can be had in a conventional electrification system. This parity is due to the incorporation of new sine wave inverters. This system allows the use of any domestic appliance.

One of the most important applications currently is the electrification of small rural nuclei with a centralized system.

To manage the energy, an electronic energy limiting device is installed in each of the homes, programmed to be able to supply the user with a daily energy, but now the equipment has to be intelligent enough to increase the assigned energy, if the state of the battery is good, or otherwise if it is bad.

The second major drawback, the power of the inverter and its performance against small loads, has been solved with the installation of several inverters in parallel.

Applications of Photovoltaic Solar Energy in Agricultural and Livestock Farms

In livestock applications, photovoltaic solar energy is used to feed pumps to provide water in cattle troughs.

Photoelectric energy is also used on farms. Farms are often far from power distribution lines. The photovoltaic panels allow in these cases to electrify the farms (lighting motors, shearers, etc.).

In certain livestock farms, photovoltaic energy is used to feed milking systems and milk cooling. Even these systems are practical for the electrification of fences.

One of the most important applications in agriculture is direct-drive water pumping. This application highlights its simplicity of installation, its null maintenance and its total automation. These systems are composed of a field of photovoltaic panels, electronic equipment and the entire system of controls and sensors for pumping.

Another application to highlight in agriculture, has been the electrification of irrigation controls and electrovalves. This electrification has allowed a better distribution and saving of water, fundamentally systems based on drip irrigation or low pressure.

SOLAR-POWERED REFRIGERATOR

Naval Special Warfare support technicians receive training on a solar-powered refrigerator.

A solar-powered refrigerator is a refrigerator which runs on energy directly provided by sun, and may include photovoltaic or solar thermal energy.

Solar-powered refrigerators are able to keep perishable goods such as meat and dairy cool in hot climates, and are used to keep much needed vaccines at their appropriate temperature to avoid spoilage.

Solar-powered refrigerators are typically used in off-the-grid locations where utility provided AC power is not available.

Technology

Solar powered refrigerators are characterized by thick insulation and the use of a DC (not AC) compressor. Traditionally solar-powered refrigerators and vaccine coolers use a combination of solar panels and lead batteries to store energy for cloudy days and at night in the absence of sunlight to keep their contents cool. These fridges are expensive and require heavy lead-acid batteries which tend to deteriorate, especially in hot climates, or are misused for other purposes. In addition, the batteries require maintenance, must be replaced approximately every three years, and must be disposed of as hazardous wastes possibly resulting in lead pollution. These problems and the resulting higher costs have been an obstacle for the use of solar powered refrigerators in developing areas.

In the mid-1990s NASA JSC began work on a solar powered refrigerator that used phase change material rather than battery to store thermal energy rather than chemical energy. The resulting technology has been commercialized and is being used for storing food products and vaccines. Solar direct-drive refrigerators don't require batteries, instead using thermal energy to solar power. These refrigerators are increasingly being used to store vaccines in remote areas.

Use

Solar-powered refrigerators and other solar appliances are commonly used by individuals living off-the-grid. They provide a means for keeping food safe and preserved while avoiding a connection to utility-provided power. Solar refrigerators are also used in cottages and camps as an alternative to absorption refrigerators, as they can be safely left running year-round. Other uses include being used to keep medical supplies at proper temperatures in remote locations, and being used to temporarily store game at hunting camps.

SOLAR LAMP

A solar lamp in Rizal Park, Philippines.

A garden solar lamp.

A solar lamp also known as solar light or solar lantern, is a lighting system composed of an LED lamp, solar panels, battery, charge controller and there may also be an inverter. The lamp operates on electricity from batteries, charged through the use of solar photovoltaic panel.

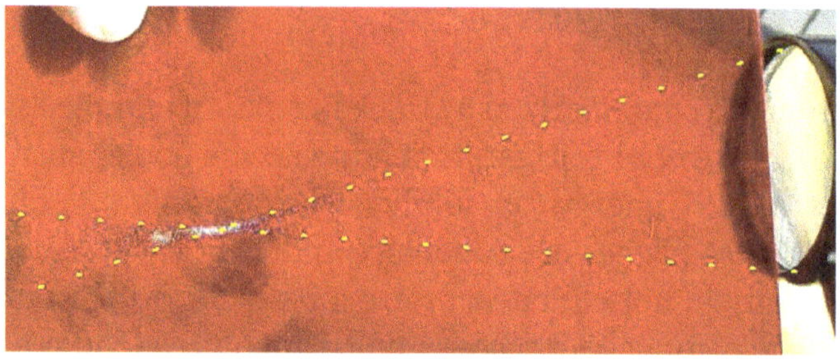

Magnifying glass.

Solar-powered household lighting can replace other light sources like candles or kerosene lamps. Solar lamps have a lower operating cost than kerosene lamps because renewable energy from the sun is free, unlike fuel. In addition, solar lamps produce no indoor air pollution unlike kerosene lamps. However, solar lamps generally have a higher initial cost, and are weather dependent.

Solar lamps for use in rural situations often have the capability of providing a supply of electricity for other devices, such as for charging cell phones. American investors have been working towards developing a RS 10 / unit solar lantern for replacement of kerosene lamps.

Components

The whole structure of solar lamp is shown in figure.

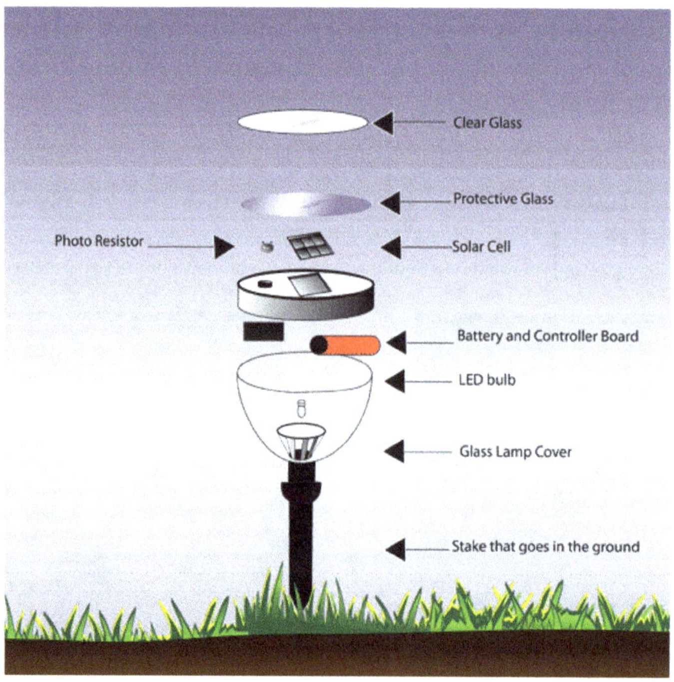

Solar Panels

Most solar panels are made out of single crystalline silicon, a semiconductor material. A solar cell has two different layers of silicon. The lower layer has less electrons and hence has a slight positive charge due to the negative charge nature of electrons. In addition, the upper layer has more electrons and has slightly negative charge. A potential barrier is created between these two layers.

When the stream of light particles called photons enter, they give up their energy to the atoms in the silicon. It promotes one electron from a covalent bond to a next energy level from upper layer to the lower layer. This promotion of an electron allows freer movement within the crystal which produces a current. More light shines through, more electrons move around hence more current flows between. This process is called photovoltaic and photoelectric effect. Photovoltaic systems directly convert sunlight into electricity.

Solar panels are made out of layers of different materials, in order of glass, encapsulate, crystalline cells, encapsulate, back sheet, junction box and lastly frame. The encapsulate keeps out moisture and contaminants which could cause problems.

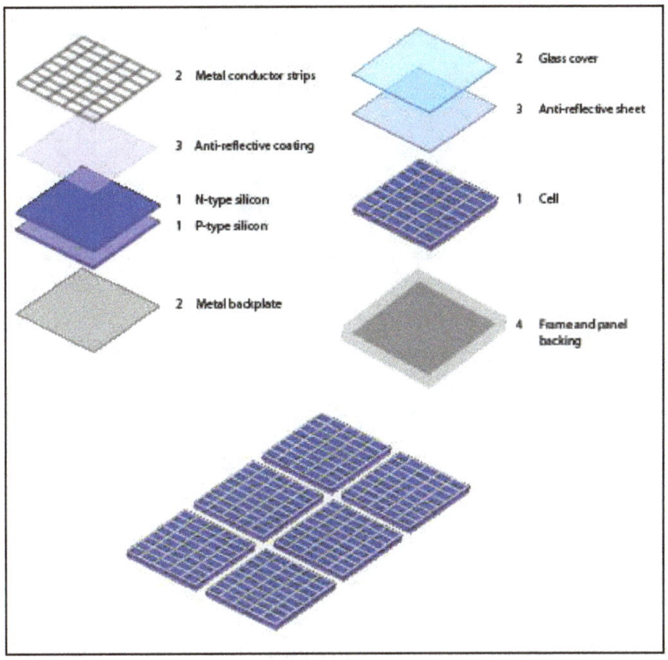

Battery

A battery is usually housed within a metal or plastic case. Inside the case are electrodes including cathodes and anodes where chemical reactions occur. A separator also exists between cathode and anode which stops the electrodes reacting together at the same time as allowing electrical charge to flow freely between the two. Lastly, the collector conducts a charge from the battery to outside.

Batteries inside solar lamps usually use gel electrolyte technology with high performance in deep discharging, in order to enable use in extreme ranges of temperature. It may also use lead-acid, nickel metal hydride, nickel cadmium, or lithium.

This part of the lamp saves up energy from the solar panel and provides power when needed at night when there is no light energy available.

In general, the efficiency of photovoltaic energy conversion is limited for physical reasons. Around 24% of solar radiation of a long wavelength is not absorbed. 33% is heat lost to surroundings, and further losses are of approximately 15-20%. Only 23% is absorbed which means a battery is a crucial part of solar lamp.

Charge Controller

This part controls the entire working systems to protect battery charge. It ensures, under any circumstances including extreme weather conditions with large temperature difference, the battery does not overcharge or over discharge and damage the battery even further.

This part also includes additional parts such as light controller, time controller, sound, temperature compensation, lighting protection, reverse polarity protection and AC transfer switches which ensure sensitive back-up loads work normally when outage occurs.

Working Principles

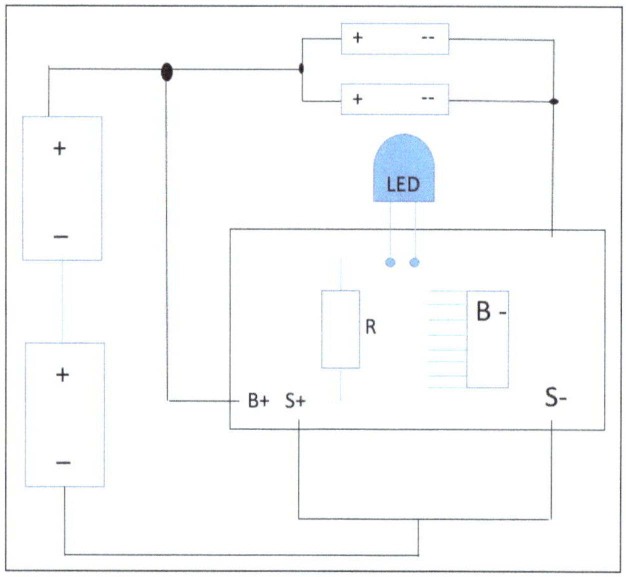

LED lights are used due to their high luminous efficiency and long life. Under the control of a DC charge controller, non-contact control automatically turns on the light at dark and switches off at daytime. It sometimes also combines with time controllers to set curtain time for it to automatically switch light on and off.

As shown in Figure, the chip includes microchip(R), B-, B+, S- and S+. S+ and S- are both connected to solar panels with wire, one of which has plus charge and the other minus charge. B- and B+ are attached to two batteries in this case. The light will be shown through the LED light when all of these are connected.

Benefits

Solar lamps can be easier for customers to install and maintain as they do not require an electricity cable. Solar lamps can benefit owners with reduced maintenance cost and costs of electricity bills. Solar lamps can also be used in areas where there is no electrical grid or remote areas that lack a reliable electricity supply. There are many stories of people with lung disease, eye deterioration, burns and sometimes even death simply because they do not have a healthy alternative to light at night. Women have felt unsafe walking to the toilet outside after dark. Babies are being delivered by midwives using only a candle, and students cannot study when the sun goes down for lack of light leading to increased illiteracy and perpetual poverty. These are the realities for over 1 billion people around the globe. Lack of lighting equates to continued poverty felt around the world. In solar energy lamp it converts light energy into electrical energy, i.e. It is convenient to our daily life.

Solar energy output is limited by weather and can be less effective if it is cloudy, wet, or winter.

Households switching to solar lamps from kerosene lamps also gain from health risk associated with kerosene emissions. Kerosene often has negative impacts on human lungs.

The use of solar energy minimizes the creation pollution indoors, where kerosene have been linked to cases of health issues. However, photovoltaic panels are made out of silicon and other toxic metals including lead that can be difficult to dispose of.

The use of solar lights improves education for students who live in households without electricity. When the nonprofit Unite-To-Light donated solar-lamps to schools a remote region of Kwa Zulu Natal in South Africa test scores and pass rates improved by over 30%. The light gives students added time to study after dark.

A 2017 experimental study in unelectrified areas of northern Bangladesh found that the use of solar lanterns decreased total household expenditure, increased children's home-study hours and increased school attendance. It did not however improve the children's educational achievement to any large extent.

Uses

Solar Street Light

These lights provide a convenient and cost-effective way to light streets at night without the need of AC electrical grids for pedestrians and drivers. They may have individual panels for each lamp of a system, or may have a large central solar panel and battery bank to power multiple lamps.

Solar LED lantern.

Rural

In rural India, solar lamps, commonly called solar lanterns, using either LEDs or CFLs, are being used to replace kerosene lamps, candles and other cheap alternatives of lighting. Especially in areas where electricity is otherwise difficult to access, solar lamps are very useful and it also improves the quality of life.

Features

Most solar lights turn on and turn off automatically by sensing outdoor light using solar panel voltage. Solar streetlights are designed to work throughout the night. Many can stay lit for more than one night if the sun is not in the sky for an extended period of time. Older models included lamps that were not fluorescent or LED. Solar lights installed in windy regions are generally equipped with flat panels to better cope with the winds.

Modern designs use wireless technology and fuzzy control theory for battery management. The street lights using this technology can operate as a network with each light having the capability of performing the turning on and off of the network.

Components

Solar street lights consist of four main parts:

Solar Panel

The solar panel is one of the most important parts of a solar street light, as the solar panel is able to convert solar energy into electricity that the lamps can use. There are two types of solar panel commonly used in solar street lights: mono-crystalline and poly-crystalline. The conversion rate of mono-crystalline solar panels are much higher than their poly-crystalline counterpart. Solar panels also vary in wattage systems.

Lighting Fixture

LEDs are usually used as the primary lighting source of modern solar street lights, as the LED will provide much higher luminosity with lower energy consumption. The energy consumption of an LED fixture is at least 50% lower than the HPS fixture counterpart which is widely used as the lighting source in traditional street lights. A lack of warm-up time in LEDs also allows for use of motion detectors for additional efficiency gains.

Rechargeable Battery

Batteries will store the electricity generated by the solar panel during the day and provide energy to the fixture during night. The life cycle of the battery is very important to the lifetime of the light and the capacity of the battery will affect the backup days of the lights. There are two types of batteries commonly used in solar-powered street lights- gel cell deep cycle batteries as well as lead acid batteries. Lithium-ion batteries are also popular these days as they are compact in size and not prone to theft due to their lack of use in other situations unlike lead acid batteries.

Pole

Strong poles are necessary to all street lights, especially to solar street lights as there are often components mounted on the top of the pole: fixtures, panels and sometimes batteries. However, in some newer designs, the PV panels and all electronics are integrated in the pole itself. Wind resistance is also a factor.

In addition, accessories do exist for these types of poles, such as a foundation cage and a battery box.

Types

Solar street light at the bus stop.

Each street light can have its own photo voltaic panel, independent of other street lights. Alternately, a number of panels can be installed as a central power source on a separate location and supply power to a number of street lights.

All-in-one type solar street lights are gaining popularity due to their compact design which incorporates all of the parts necessary in a compact manner including the battery.

The city of Las Vegas, Nevada was the first city in the world that tested new EnGoPlanet Solar Street lights which are coupled with kinetic tiles that produce electricity when people walk over them.

Advantages

- Solar street lights are independent of the utility grid. Hence, the operation costs are minimized.
- Solar street lights require much less maintenance compared to conventional street lights.
- Since external wires are eliminated, risk of accidents are minimized.
- Electricity produced from solar panels is non-pollutive.
- Separate parts of a solar panel system can easily be transported.
- Energy costs can be saved.

Disadvantages

- Initial investment is higher compared to conventional street lights.
- Risk of theft is higher as equipment costs are comparatively higher.
- Snow or dust, combined with moisture can accumulate on horizontal PV-panels and reduce or even stop energy production.
- Rechargeable batteries will need to be replaced several times over the lifetime of the fixtures adding to the total lifetime cost of the light.

The charge and discharge cycles of the battery are also very important considering the overall cost of the project.

SOLAR-POWERED FAN

A solar fan with externally mounted solar panels.

A solar fan is a mechanical fan powered by solar panels. The solar panels are either mounted on the device or are installed independently. Solar fans mostly do not require secondary power sources other than solar power, as most of them are used for cooling purposes during day time. Some types are also used for heating purposes. It runs the fastest when it is the hottest outside providing savings on air conditioning costs.

Applications

Attic heat is caused by the sun baking down on the roof surface. This sunlight causes the roofing material to heat up, thus transferring the heat into the attic via thermal transfer. The buildup of heat is generally dissipated by normal convection, i.e.: as the air is heated it expands and rises, flowing out of the attic through vents placed at or near the peak, while fresh air is pulled into the attic from intake vents that are generally in the eave (soffit) area. However, most roof vents are inadequate at expelling the heat and require the use of a powered vent fan to remove heat effectively. A solar powered attic fan is an effective way to remove attic heat using only the power of the sun to run the fan.

Solar-powered Attic Fan - Roof Mounted

A typical roof-mounted solar attic fan will be effective in removing excessive heat from the attic. When mounted near the peak the solar fan forces hot air out while pulling fresh air in via eave/ soffit vents. This exchange of air keeps the attic space from overheating. Solar attic fans not only lower the cooling costs in your home, they also extend the life of your shingle roof.

Solar-powered Attic Fan - Gable Mounted

It is similar to the roof mounted solar fan, except that its installation is a little different. Most attic fans are installed on the roof surface, whereas gable fans are installed in the gables, inside the attic and behind an existing louvered gable vent as the name suggests. They are mostly installed in places like greenhouse, cabin, RV's garages or barns and other storage places where heat and moisture are usually present. These types of fans are generally installed behind intake grills to provide ventilation. Gable Solar Fans are usually installed high on the gable close to the ridge and are also coupled with soffit or roof vents for well-balanced intake and exhaust air avenues. The airflow depends upon the level of sunlight when there is more sunlight, the airflow will be more resulting in enhanced cooling. It is also capable of lowering the air conditioning costs by cooling the attic.

Since the only source of energy is from the solar panel for a solar fan. It is important to have a larger solar panel to get the better performance from the solar fan. Amtrak Solar makes one of the most powerful solar fans in United States of America.

In 2007, solar fans were installed on the terminal roofs of Honolulu International Airport, in addition to the mansion of Indiana governor. In 2009 the Indiana State Senate passed a bill giving a tax deduction to homeowners with solar fans.

Advantages

- A solar fan is more environmentally friendly.
- Though initially the installation may be costlier, in the long term it would turn out to be cheaper because it does not use power from the utility grid and may provide savings up to 30 percent on air conditioning costs.
- Risk of electric accidents are eliminated as there are no electric cords attached to the solar fan.
- A cordless appliance offers much more mobility compared to conventional fans.

SOLAR VEHICLE

A solar vehicle is an electric vehicle powered completely or significantly by direct solar energy. Usually, photovoltaic (PV) cells contained in solar panels convert the sun's energy directly into electric energy. The term "solar vehicle" usually implies that solar energy is used to power all or part of a vehicle's propulsion. Solar power may be also used to provide power for communications or controls or other auxiliary functions.

U.S. Secretary of State John Kerry admires a solar-powered car built by members of the Tomodachi Initiative youth engagement program.

Solar vehicles are not sold as practical day-to-day transportation devices at present, but are primarily demonstration vehicles and engineering exercises, often sponsored by government agencies. However, indirectly solar-charged vehicles are widespread and solar boats are available commercially.

Land

Solar Cars

Solar cars depend on PV cells to convert sunlight into electricity to drive electric motors. Unlike solar thermal energy which converts solar energy to heat, PV cells directly convert sunlight into electricity.

The design of a solar car is severely limited by the amount of energy input into the car. Solar cars are built for solar car races and also for public use List of prototype solar-powered cars. Even the best solar cells can only collect limited power and energy over the area of a car's surface. This limits solar cars to ultralight composite bodies to save weight. Solar cars lack the safety and convenience features of conventional vehicles. The first solar family car was built in 2013 by students in the Netherlands. This vehicle is capable of 550 miles on one charge during sunlight. It weighs 850 pounds and has a 1.5kw solar array. Solar vehicles must be light and efficient. 3,000 pound or even 2,000 pound vehicles are less practical. Stella Lux, the predecessor to Stella, broke a record with a 932 mile single charge range. The Dutch are trying to commercialize this technology. During racing Stella Lux is capable of 700 miles during daylight. At 45 mph Stella Lux has infinite range. This is again due to high efficiency including a Coefficient of drag of .16. The average family who never drive more than 200 miles a day would never need to charge from the mains. They would only plug in if they wanted to return energy to the grid. Solar cars are often fitted with gauges and/or wireless telemetry, to carefully monitor the car's energy consumption, solar energy capture and other parameters. Wireless telemetry is typically preferred as it frees the driver to concentrate on driving, which can be dangerous in such a small, lightweight car. The Solar Electric Vehicle system was designed and engineered as an easy to install (2 to 3 hours) integrated accessory system with a custom molded low profile solar module, supplemental battery pack and a proven charge controlling system.

As an alternative, a battery-powered electric vehicle may use a solar array to recharge; the array may be connected to the general electrical distribution grid.

Solar Buses

Solar buses are propulsed by solar energy, all or part of which is collected from stationary solar panel installations. The Tindo bus is a 100% solar bus that operates as free public transport service in Adelaide City as an initiative of the City Council. Bus services which use electric buses that are partially powered by solar panels installed on the bus roof, intended to reduce energy consumption and to prolong the life cycle of the rechargable battery of the electric bus, have been put in place in China.

Solar buses are to be distinguished from conventional buses in which electric functions of the bus such as lighting, heating or air-conditioning, but not the propulsion itself, are fed by solar energy. Such systems are more widespread as they allow bus companies to meet specific regulations, for example the anti-idling laws that are in force in several of the US states, and can be retrofitted to existing vehicle batteries without changing the conventional engine.

Single-track Vehicles

The first solar "cars" were actually tricycles or Quadracycles built with bicycle technology. These were called solarmobiles at the first solar race, the Tour de Sol in Switzerland in 1985. With 72 participants, half used solar power exclusively while the other half used solar-human-powered hybrids. A few true solar bicycles were built, either with a large solar roof, a small rear panel, or a trailer with a solar panel. Later more practical solar bicycles were built with foldable panels to be set up only during parking. Even later the panels were left at home, feeding into the electric mains, and the bicycles charged from the mains. Today highly developed electric bicycles are available and these use so little power that it costs little to buy the equivalent amount of solar electricity. The "solar" has evolved from actual hardware to an indirect accounting system. The same system also works for electric motorcycles, which were also first developed for the Tour de Sol.

Applications

The Venturi Astrolab in 2006 was the world's first commercial electro-solar hybrid car, and was originally due to be released in January 2008.

In May 2007 a partnership of Canadian companies led by Hymotion altered a Toyota Prius to use solar cells to generate up to 240 watts of electrical power in full sunshine. This is reported as permitting up to 15 km extra range on a sunny summer day while using only the electric motors.

An inventor from Michigan, USA built a street legal, licensed, insured, solar charged electric scooter in 2005. It had a top speed controlled at a bit over 30 mph, and used fold-out solar panels to charge the batteries while parked.

Auxiliary Power

Nuna 3 PV powered car.

Photovoltaic modules are used commercially as auxiliary power units on passenger cars in order to ventilate the car, reducing the temperature of the passenger compartment while it is parked in the sun. Vehicles such as the 2010 Prius, Aptera 2, Audi A8, and Mazda 929 have had solar sunroof options for ventilation purposes.

The area of photovoltaic modules required to power a car with conventional design is too large to be carried on board. A prototype car and trailer has been built Solar Taxi. According to the website, it is capable of 100 km/day using 6m^2 of standard crystalline silicon cells. Electricity is stored using a nickel/salt battery. A stationary system such as a rooftop solar panel, however, can be used to charge conventional electric vehicles.

It is also possible to use solar panels to extend the range of a hybrid or electric car, as incorporated in the Fisker Karma, available as an option on the Chevy Volt, on the hood and roof of "Destiny 2000" modifications of Pontiac Fieros, Italdesign Quaranta, Free Drive EV Solar Bug, and numerous other electric vehicles, both concept and production. In May 2007 a partnership of Canadian companies led by Hymotion added PV cells to a Toyota Prius to extend the range. SEV claims 20 miles per day from their combined 215W module mounted on the car roof and an additional 3kWh battery.

On 9 June 2008, the German and French Presidents announced a plan to offer a credit of 6-8g/km of CO_2 emissions for cars fitted with technologies "not yet taken into consideration during the standard measuring cycle of the emissions of a car". This has given rise to speculation that photovoltaic panels might be widely adopted on autos in the near future

It is also technically possible to use photovoltaic technology, (specifically thermophotovoltaic (TPV) technology) to provide motive power for a car. Fuel is used to heat an emitter. The infrared radiation generated is converted to electricity by a low band gap PV cell (e.g. GaSb). A prototype TPV hybrid car was even built. The "Viking 29" was the World's first thermophotovoltaic (TPV) powered automobile, designed and built by the Vehicle Research Institute (VRI) at Western

Washington University. Efficiency would need to be increased and cost decreased to make TPV competitive with fuel cells or internal combustion engines.

Personal Rapid Transit

Several personal rapid transit (PRT) concepts incorporate photovoltaic panels.

Rail

Railway presents a low rolling resistance option that would be beneficial of planned journeys and stops. PV panels were tested as APUs on Italian rolling stock under EU project.PVTRAIN. Direct feed to a DC grids avoids losses through DC to AC conversion. DC grids are only to be found in electric powered transport: railways, trams and trolleybuses. Conversion of DC from PV panels to grid alternating current (AC) was estimated to cause around 3% of the electricity being wasted.

JPods PRT concept with photovoltaic panels above guideways.

PVTrain concluded that the most interest for PV in rail transport was on freight cars where on board electrical power would allow new functionality:

- GPS or other positioning devices, so as to improve its use in fleet management and efficiency.
- Electric locks, a video monitor and remote control system for cars with sliding doors, so as to reduce the risk of robbery for valuable goods.
- ABS brakes, which would raise the maximum velocity of freight cars to 160 km/h, improving productivity.

The Kismaros – Királyrét narrow-gauge line near Budapest has built a solar powered railcar called 'Vili'. With a maximum speed of 25 km/h, 'Vili' is driven by two 7 kW motors capable of regenerative braking and powered by 9.9m2 of PV panels. Electricity is stored in on-board batteries. In addition to on-board solar panels, there is the possibility to use stationary (off-board) panels to generate electricity specifically for use in transport.

A few pilot projects have also been built in the framework of the "Heliotram" project, such as the tram depots in Hannover Leinhausen and Geneva (Bachet de Pesay). The 150 kW_p Geneva site injected 600 V DC directly into the tram/trolleybus electricity network provided about 1% of the electricity used by the Geneva transport network at its opening in 1999. On December 16, 2017 a fully solar-powered train was launched in New South Wales, Australia. The train is powered using onboard solar panels and onboard rechargeable batteries. It holds a capacity for 100 seated passenger for a 3 km journey.

Recently Imperial College London and the environmental charity 10:10 have announced the Renewable Traction Power project to investigate using track-side solar panels to power trains. Meanwhile, Indian railways announced their intention to use on board PV to run air conditioning systems in railway coaches.. Also, Indian Railways announced it is to conduct a trial run by the end of May 2016. It hopes that an average of 90,800 liters of diesel per train will be saved on an annual basis, which in turn results in reduction of 239 tonnes of CO_2.

Water

PlanetSolar, the world's largest solar-powered boat and the first ever solar electric vehicle to circumnavigate the globe.

Solar powered boats have mainly been limited to rivers and canals, but in 2007 an experimental 14m catamaran, the Sun21 sailed the Atlantic from Seville to Miami, and from there to New York. It was the first crossing of the Atlantic powered only by solar.

Japan's biggest shipping line Nippon Yusen KK and Nippon Oil Corporation said solar panels capable of generating 40 kilowatts of electricity would be placed on top of a 60,213 ton car carrier ship to be used by Toyota Motor Corporation.

In 2010, the Tûranor PlanetSolar, a 30 metre long, 15.2 metre wide catamaran yacht powered by 470 square metres of solar panels, was unveiled. It is, so far, the largest solar-powered boat ever built. In 2012, PlanetSolar became the first ever solar electric vehicle to circumnavigate the globe.

Various demonstration systems have been made. Curiously, none yet takes advantage of the huge power gain that water cooling would bring.

The low power density of current solar panels limits the use of solar propelled vessels, however boats that use sails (which do not generate electricity unlike combustion engines) rely on battery power for electrical appliances (such as refrigeration, lighting and communications). Here solar panels have become popular for recharging batteries as they do not create noise, require fuel and often can be seamlessly added to existing deck space.

Air

The Swiss solar-powered aircraft Solar Impulse completed a circumnavigation of the world in 2016.

Gossamer Penguin.

Solar ships can refer to solar powered airships or hybrid airships.

There is considerable military interest in unmanned aerial vehicles (UAVs); solar power would enable these to stay aloft for months, becoming a much cheaper means of doing some tasks done today by satellites. In September 2007, the first successful flight for 48h under constant power of a UAV was reported. This is likely to be the first commercial use for photovoltaics in flight.

Many demonstration solar aircraft have been built, some of the best known by AeroVironment.

Manned Solar Aircraft

- Gossamer Penguin,
- Solar Challenger: This aircraft flew 163 miles (262 km) from Paris, France to England on solar power.
- Sunseeker.
- Solar Impulse: Two single-seat aircraft, the second of which circumnavigated the Earth. The first aircraft completed a 26-hour test flight in Switzerland on 8-9 July 2010. The aircraft was flown to a height of nearly 28,000 feet (8,500 meters) by Andre Borschberg. It flew overnight using battery power. The second aircraft, slightly larger and more powerful, took off from Abu Dhabi in 2015, flew towards India and then eastward across Asia. However, after experiencing battery overheating, it was forced to halt in Hawaii over the winter. In April 2016, it resumed its journey, and completed its circumnavigation of the globe, returning to Abu Dhabi on 26 July 2016.
- SolarStratos: Swiss stratospheric 2-seater solar plane aims to climb into space.

Hybrid Airships

An Australian-based company is working on a project to develop an air crane called the SkyLifter, a "vertical pick-up and delivery aircraft" being capable of lifting up to 150 tons.

A Canadian start-up, Solar Ship Inc, is developing solar powered hybrid airships that can run on solar power alone. The idea is to create a viable platform that can travel anywhere in the world delivering cold medical supplies and other necessitates to locations in Africa and Northern Canada without needing any kind of fuel or infrastructure. The hope is that technology developments in solar cells and the large surface area provided by the hybrid airship are enough to make a practical solar powered aircraft. Some key features of the Solarship are that it can fly on aerodynamic lift alone without any lifting gas, and the solar cells along with the large volume of the envelope allow the hybrid airship to be reconfigured into a mobile shelter that can recharge batteries and other equipment.

The Hunt GravityPlane is a proposed gravity-powered glider by Hunt Aviation in the USA. It also has aerofoil wings, improving its lift-drag ratio and making it more efficient. The GravityPlane requires a large size in order to obtain a large enough volume-to-weight ratio to support this wing structure, and no example has yet been built. Unlike a powered glider, the GravityPlane does not consume power during the climbing phase of flight. It does however consume power at the points where it changes its buoyancy between positive and negative values. Hunt claim that this can nevertheless improve the energy efficiency of the craft, similar to the improved energy efficiency of underwater gliders over conventional methods of propulsion. Hunt suggest that the low power consumption should allow the craft to harvest sufficient energy to stay aloft indefinitely. The conventional approach to this requirement is the use of solar panels in a solar-powered aircraft. Hunt has proposed two alternative approaches. One is to use a wind turbine and harvest energy from the airflow generated by the gliding motion, the other is a thermal cycle to extract energy from the differences in air temperature at different altitudes.

Unmanned Aerial Vehicles

- Pathfinder and Pathfinder-Plus: This UAV demonstrated that an airplane could stay aloft for an extended period of time fueled purely by solar power.
- Helios: Derived from the Pathfinder-Plus, this solar cell and fuel cell powered UAV set a world record for flight at 96,863 feet (29,524 m).
- Qinetiq Zephyr: Built by Qinetiq, this UAV set the unofficial world record for longest duration unmanned flight at over 82 hours on 31 July 2008. Just 15 days after the Solar Impulse flight mentioned above, on 23 July 2010 the Zephyr, a lightweight unmanned aerial vehicle engineered by the United Kingdom defence firm QinetiQ, claimed the endurance record for an unmanned aerial vehicle. It flew in the skies of Arizona for over two weeks (336 hours). It has also soared to over 70,700 feet (21.5 km).
- China's designed and manufactured UAV successfully reached an altitude of 20,000 meters during a test flight in the country's northwest regions. Named "Caihong" (CH), or "Rainbow" in English, it was developed by a research team from CASC.

Space

Solar Powered Spacecraft

PV on the International Space Station.

Solar energy is often used to supply power for satellites and spacecraft operating in the inner solar system since it can supply energy for a long time without excess fuel mass. A Communications satellite contains multiple radio transmitters which operate continually during its life. It would be uneconomic to operate such a vehicle (which may be on-orbit for years) from primary batteries or fuel cells, and refuelling in orbit is not practical. Solar power is not generally used to adjust the satellite's position, however, and the useful life of a communications satellite will be limited by the on-board station-keeping fuel supply.

Solar Propelled Spacecraft

A few spacecraft operating within the orbit of Mars have used solar power as an energy source for their propulsion system.

All current solar powered spacecraft use solar panels in conjunction with electric propulsion, typically ion drives as this gives a very high exhaust velocity, and reduces the propellant over that of a rocket by more than a factor of ten. Since propellant is usually the biggest mass on many spacecraft, this reduces launch costs.

Other proposals for solar spacecraft include solar thermal heating of propellant, typically hydrogen or sometimes water is proposed. An electrodynamic tether can be used to change a satellite's orientation or adjust its orbit.

Another concept for solar propulsion in space is the light sail; this doesn't require conversion of light to electrical energy, instead relying directly on the tiny but persistent radiation pressure of light.

Planetary Exploration

Perhaps the most successful solar-propelled vehicles have been the "rovers" used to explore surfaces of the Moon and Mars. The 1977 Lunokhod programme and the 1997 Mars Pathfinder used solar power to propel remote controlled vehicles. The operating life of these rovers far exceeded the limits of endurance that would have been imposed, had they been operated on conventional fuels.

Electric Vehicle with Solar Assist

A Swiss project, called "Solartaxi", has circumnavigated the world. This is the first time in history an electric vehicle (not self sufficient solar vehicle) has gone around the world, covering 50000 km in 18 months and crossing 40 countries. It is a road-worthy electric vehicle hauling a trailer with solar panels, carrying a 6 m^2 sized solar array. The Solartaxi has Zebra batteries, which permit a range of 400 km without recharging. The car can also run for 200 km without the trailer. Its maximum speed is 90 km/h. The car weighs 500 kg and the trailer weighs 200 kg. According to initiator and tour director Louis Palmer, the car in mass production could be produced for 16000 Euro. Solartaxi has toured the World from July 2007 till December 2008 to show that solutions to stop global warming are available and to encourage people in pursuing alternatives to fossil fuel. Palmer suggests the most economical location for solar panels for an electric car is on building rooftops though, likening it to putting money into a bank in one location and withdrawing it in another.

Solar Electrical Vehicles is adding convex solar cells to the roof of hybrid electric vehicles.

Plug-in Hybrid and Solar Vehicles

Louis Palmer standing in the Solartaxi.

An interesting variant of the electric vehicle is the triple hybrid vehicle—the PHEV that has solar panels as well to assist.

The 2010 Toyota Prius model has an option to mount solar panels on the roof. They power a ventilation system while parked to help provide cooling. There are many applications of photovoltaics in transport either for motive power or as auxiliary power units, particularly where fuel, maintenance, emissions or noise requirements preclude internal combustion engines or fuel cells. Due to the limited area available on each vehicle either speed or range or both are limited when used for motive power.

Limitations

PV used for auxiliary power on a yacht.

There are limits to using photovoltaic (PV) cells for vehicles:

- Power density: Power from a solar array is limited by the size of the vehicle and area that can be exposed to sunlight. This can also be overcome by adding a flatbed and connecting it to the car and this gives more area for panels for powering the car. While energy can be accumulated in batteries to lower peak demand on the array and provide operation in sunless conditions, the battery adds weight and cost to the vehicle. The power limit can be mitigated by use of conventional electric cars supplied by solar (or other) power, recharging from the electrical grid.
- Cost: While sunlight is free, the creation of PV cells to capture that sunlight is expensive. Costs for solar panels are steadily declining (22% cost reduction per doubling of production volume).
- Design considerations: Even though sunlight has no lifespan, PV cells do. The lifetime of a solar module is approximately 30 years. Standard photovoltaics often come with a warranty of 90% (from nominal power) after 10 years and 80% after 25 years. Mobile applications are unlikely to require lifetimes as long as building integrated PV and solar parks. Current PV panels are mostly designed for stationary installations. However, to be successful in mobile applications, PV panels need to be designed to withstand vibrations. Also, solar panels, especially those incorporating glass, have significant weight. In order for its addition to be of value, a solar panel must provide energy equivalent to or greater than the energy consumed to propel its weight.

SOLAR PANELS ON SPACECRAFT

A solar panel array of the International Space Station.

Spacecraft operating in the inner Solar System usually rely on the use of photovoltaic solar panels to derive electricity from sunlight. Outside the orbit of Jupiter, solar radiation is too weak to produce sufficient power within current solar technology and spacecraft mass limitations, so radioisotope thermoelectric generators (RTGs) are instead used as a power source.

Uses

The solar panels on the SMM satellite provided electrical power. Here it is being captured by an astronaut in a mobile space-suit that runs on chemical battery power.

Solar panels on spacecraft supply power for two main uses:

- Power to run the sensors, active heating, cooling and telemetry.
- Power for spacecraft propulsion – electric propulsion, sometimes called solar-electric propulsion.

For both uses, a key figure of merit of the solar panels is the specific power (watts generated divided by solar array mass), which indicates on a relative basis how much power one array will generate for a given launch mass relative to another. Another key metric is stowed packing efficiency (deployed watts produced divided by stowed volume), which indicates how easily the array will fit into a launch vehicle. Yet another key metric is cost (dollars per watt).

To increase the specific power, typical solar panels on spacecraft use close-packed solar cell rectangles that cover nearly 100% of the Sun-visible area of the solar panels, rather than the solar wafer circles which, even though close-packed, cover about 90% of the Sun-visible area of typical solar panels on Earth. However, some solar panels on spacecraft have solar cells that cover only 30% of the Sun-visible area.

Implementation

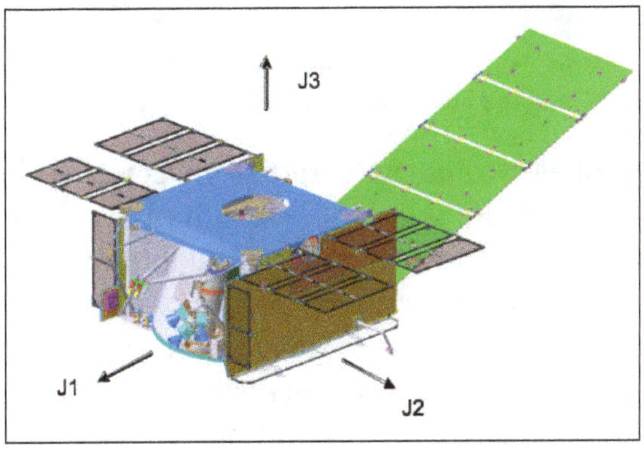

Diagram of the Spacecraft Bus on the planned James Webb Space Telescope, which is powered by solar panels (coloured green in this 3/4 view). Note that shorter light purple extensions are radiator shades not solar panels.

Solar panels need to have a lot of surface area that can be pointed towards the Sun as the spacecraft moves. More exposed surface area means more electricity can be converted from light energy from the Sun. Since spacecraft have to be small, this limits the amount of power that can be produced.

All electrical circuits generate waste heat; in addition, solar arrays act as optical and thermal as well as electrical collectors. Heat must be radiated from their surfaces. High-power spacecraft may have solar arrays that compete with the active payload itself for thermal dissipation. The innermost panel of arrays may be "blank" to reduce the overlap of views to space. Such spacecraft include the higher-power communications satellites (e.g., later-generation TDRS) and Venus Express, not high-powered but closer to the Sun.

Spacecraft are built so that the solar panels can be pivoted as the spacecraft moves. Thus, they can always stay in the direct path of the light rays no matter how the spacecraft is pointed. Spacecraft are usually designed with solar panels that can always be pointed at the Sun, even as the rest of the body of the spacecraft moves around, much as a tank turret can be aimed independently of where the tank is going. A tracking mechanism is often incorporated into the solar arrays to keep the array pointed towards the sun.

Sometimes, satellite operators purposefully orient the solar panels to "off point," or out of direct alignment from the Sun. This happens if the batteries are completely charged and the amount of electricity needed is lower than the amount of electricity made; off-pointing is also sometimes used on the International Space Station for orbital drag reduction.

Ionizing Radiation Issues and Mitigation

Juno is the second spacecraft to orbit Jupiter and the first solar-powered craft to do so.

Space contains varying levels of electromagnetic radiation as well as ionizing radiation. Solar Energetic Particles and Galactic Cosmic Rays are the two radiation sources of interest. The Van Allen belts contain mostly protons and electrons, while GCR contains ionized nucleons ranging from Helium to Uranium. Solar panels will experience efficiency degradation over time as a result of both kinds of radiation. With borosilicate glass panel coverings, this may be between 5-10% efficiency loss per year. Other glass coverings, such as fused silica and lead glasses, may reduce this efficiency loss to less than 1% per year. The degradation rate is a function of the differential flux spectrum and the total ionizing dose.

Types of Solar Cells Typically used

Gallium arsenide-based solar cells are typically favored over crystalline silicon in industry because they have a higher efficiency and degrade more slowly than silicon in the radiation present in space. The most efficient solar cells currently in production are multi-junction photovoltaic cells. These use a combination of several layers of gallium arsenide, indium gallium phosphide, and germanium to capture more energy from the solar spectrum. Leading edge multi-junction cells are capable of exceeding 38.8% under non-concentrated AM1.5G illumination and 46% using concentrated AM1.5G illumination.

Spacecraft that have used Solar Power

To date, solar power, other than for propulsion, has been practical for spacecraft operating no farther from the Sun than the orbit of Jupiter. For example, Juno, Magellan, Mars Global Surveyor, and Mars Observer used solar power as does the Earth-orbiting, Hubble Space Telescope. The Rosetta space probe, launched 2 March 2004, used its 64 square metres (690 sq ft) of solar panels as far as the orbit of Jupiter (5.25 AU); previously the furthest use was the Stardust spacecraft at 2 AU. Solar power for propulsion was also used on the European lunar mission SMART-1 with a Hall effect thruster.

The Juno mission, launched in 2011, is the first mission to Jupiter (arrived at Jupiter on July 4, 2016) to use solar panels instead of the traditional RTGs that are used by previous outer Solar System missions, making it the furthest spacecraft to use solar panels to date. It has 72 square metres (780 sq ft) of panels.

Solar panels extended out from the Apollo Telescope Mount, power solar observatory instruments on the Skylab station, which also had an additional array on the main spacecraft.

Another spacecraft of interest is *Dawn* which went into orbit around 4 Vesta in 2011. It used ion thrusters to get to Ceres.

The potential for solar powered spacecraft beyond Jupiter has been studied.

The International Space Station also uses solar arrays to power everything on the station. The 262,400 solar cells cover around 27,000 square feet (2,500 m^2) of space. There are four sets of solar arrays that power the station and the fourth set of arrays were installed in March 2009. 84 to 120 kilowatts of electricity can be generated from these solar arrays.

SOLAR-POWERED PUMP

A windpump replaced by a solar powered pump at a water hole in the Augrabies Falls National Park.

This solar water pump up to 5 hp is useful for farmers.

A solar-powered pump is a pump running on electricity generated by photovoltaic panels or the radiated thermal energy available from collected sunlight as opposed to grid electricity or diesel run water pumps. The operation of solar powered pumps is more economical mainly due to the lower operation and maintenance costs and has less environmental impact than pumps powered by an internal combustion engine (ICE). Solar pumps are useful where grid electricity is unavailable and alternative sources (in particular wind) do not provide sufficient energy.

Components

A photovoltaic solar powered pump system has three parts:

- Solar Panels.
- The Controller.
- The pump.

The solar panels make up most (up to 80%) of the systems cost. The size of the PV-system is directly dependent on the size of the pump, the amount of water that is required (m^3/d) and the solar irradiance available.

The purpose of the controller is twofold. Firstly, it matches the output power that the pump receives with the input power available from the solar panels. Secondly, a controller usually provides a low voltage protection, whereby the system is switched off, if the voltage is too low or too high for the operating voltage range of the pump. This increases the lifetime of the pump thus reducing the need for maintenance. Other ancillary functions include automatically shutting down the system when water source level is low or when the storage tank is full, regulating water output pressure, blending power input between the solar panels and an alternate power source such as the grid or a petrol generator, and remotely monitoring and managing the system through an online portal offered as a cloud service by the manufacturer.

Voltage of the solar pump motors can be AC (alternating current) or DC (direct current). Direct current motors are used for small to medium applications up to about 4 kW rating, and are suitable for applications such as garden fountains, landscaping, drinking water for livestock, or small irrigation projects. Since DC systems tend to have overall higher efficiency levels than AC pumps of a similar size, the costs are reduced as smaller solar panels can be used.

Finally, if an alternating current solar pump is used, an inverter is necessary that changes the direct current from the solar panels into alternating current for the pump. The supported power range of inverters extends from 0.15 to 55 kW and can be used for larger irrigation systems. However, the panel and inverters must be sized accordingly to accommodate the inrush characteristic of an AC motor. To aid in proper sizing, leading manufacturers provide proprietary sizing software tested by third party certifying companies. The sizing software may include the projected monthly water output which varies due to seasonal change in insolation.

Water Pumping

Solar powered water pumps can deliver drinking water as well as water for livestock or irrigation purposes. Solar water pumps may be especially useful in small scale or community based irrigation, as large scale irrigation requires large volumes of water that in turn require a large solar PV array. As the water may only be required during some parts of the year, a large PV array would provide excess energy that is not necessarily required, thus making the system inefficient.

Solar PV water pumping systems are used for irrigation and drinking water in India. The majority of the pumps are fitted with a 2000 watt - 3,700 watt motor that receives energy from a 4,800 Wp PV array. The 5hp systems can deliver about 124,000 liters of water/day from a total of 50 meters setoff head and 70 meters dynamic head. By 30 August 2016, a total of 1,20,000 solar PV water pumping systems have been installed in India and many other places around the world. Energy storage in the form of water storage is better than energy storage in the form of batteries for solar water pumps because there is no intermediary transformation of one form of energy to another. The most common pump mechanics used are centrifugal pumps, multistage pumps, borehole pumps, and helical pumps. Important scientific concepts of fluid dynamics like pressure vs. head, pump heads, pump curves, system curves, and net suction head are really important for the successful deployment and design of solar powered pumps.

Oil and Gas

In order to combat negative publicity related to the environmental impacts of fossil fuels, including fracking, the industry is embracing solar powered pumping systems. Many oil and gas wells require the accurate injection (metering) of various chemicals under pressure to sustain their operation and to improve extraction rates. Historically, these chemical injection pumps (CIP) have been driven by gas reciprocating motors utilizing the pressure of the well's gas and exhausting the raw gas into the atmosphere. Solar powered electrical pumps (solar CIP) can reduce these greenhouse gas emissions. Solar arrays (photovoltaic cells) not only provide a sustainable power source for the CIPs but can also provide an electric source to run remote SCADA type diagnostics with remote control and satellite/cell communications from very remote locations to a desktop or notebook monitoring computer.

Stirling Engine

Instead of generating electricity to turn a motor, sunlight can be concentrated on the heat exchanger of a Stirling engine and used to drive a pump mechanically. This dispenses with the cost of solar panels and electric equipment. In some cases the Stirling engine may be suitable for local fabrication, eliminating the difficulty of importing equipment. One form of Stirling engine is the fluidyne engine which operates directly on the pumped fluid as a piston. Fluidyne solar pumps have been studied since 1987 At least one manufacturer has conducted tests with a Stirling solar powered pump.

SOLAR-POWERED WATCH

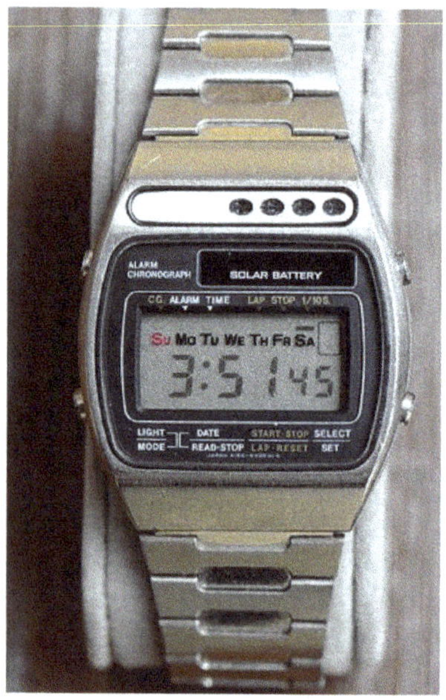

LCD Solar Alarm Chronograph.

A solar-powered watch or light-powered watch is a watch that is powered entirely or partly by a solar cell.

Some of the early solar watches of the 1970s had innovative and unique designs to accommodate the array of photovoltaic solar cells needed to power them (Synchronar, Nepro, Sicura and some models by Cristalonic, Alba, Rhythm, Seiko and Citizen). In the 1990s, Citizen started to sell light-powered watches under the Eco-Drive series name. Since their introduction, photovoltaic devices have greatly improved their efficiency and thereby their capacity. Watchmakers have developed their technology such that solar-powered watches are now a major part of their range of watches. Several other watch manufacturers offer similar watches or are developing such technology. These other watch manufacturers include (amongst others) Junghans, Casio, Seiko, and Orient.

Typically, sunlight and artificial light are absorbed by a solar panel behind the crystal. The dial is either on a layer above or actually on the solar panel. This solar panel converts the light into electrical energy to power the watch. The watch will usually store energy in a rechargeable cell to power itself during the night or when covered such as a wearer's clothing (e.g., sleeve). Citizen's watches use lithium-ion batteries to store sufficient energy to power the watch for several months/years without light exposure, by allowing the watch to enter a power-saving or hibernation mode during which the seconds hand stops until the watch is re-exposed to light. Not all have a power-save mode, yet will still hold a charge for typically six months, as with uncomplicated (date only) analog versions made by Citizen.

A modern Casio solar-powered watch.

Inexpensive solar-powered watches were first sold in the 1980s and were popular amongst children, often featuring famous fictional characters such as Transformers or G.I. Joe.

SOLAR-POWERED FLASHLIGHT

A solar flashlight with built-in radio.

Solar powered flashlights or solar powered torches are flashlights powered by solar energy stored in rechargeable batteries. Most of these flashlights use light-emitting diodes lamps since they have lower energy consumption compared to incandescent light bulbs.

Features

Solar powered flashlights vary in features and capabilities. A typical solar flashlight can give useful levels of illumination on objects up to 50 metres away, and beam may be visible for much longer distances. The solar cells used for battery charging have an indefinite life expectancy. A solar powered flashlight may give several hours of light after being charged during the day. These flashlights may be designed to be impact resistant, weatherproof, and to float.

Other models include features such as a solar cell phone charger, an AM/FM radio, or a siren to call for help in an emergency. Some models include a hand crank dynamo for charging at night. An experimental solar flashlight the size of a credit card features a white LED powered by 16 solar cells.

Comparison with Primary Battery Flashlights

Solar powered lights need not be manually recharged as in the case of conventional battery operated flashlights. Pollution from discarded batteries used in normal flashlights is eliminated. Solar powered flashlights can be used in remote locations where it is impossible to find any source of electricity except for solar power.

A built-in solar cell array in a hand-held product has a very small capacity. The light output and run time of the solar flashlight are limited by the amount of energy that can be absorbed in a day. Flashlights are usually stored indoors or otherwise out of direct sunlight; consumers may find it inconvenient to remember to place the flashlight in full sunlight for several hours before use. Rechargeable cells self-discharge, so unless the flashlight is stored in a sunny location, the light may not function in an emergency. Rechargeable cells have a finite charge/discharge cycle life, and when the cells wear out it may be difficult for the consumer to replace them. Some of these disadvantages can be mitigated by separating the solar cell array from the flashlight, but then the simplicity of automatic charging is lost.

Significance

Solar powered flashlights have been distributed to countries where there is either no or erratic power supply, helping people feel safer leaving their homes at night, and giving children the opportunity to study after sunset. A dual purpose solar flashlight, which also functions as a lamp has helped families in such countries spend more time together, surgeries to be performed during power outages and people to travel in the dark under a halo of light.

SOLAR AIR CONDITIONING

A solar air conditioning system is simply a system of cooling and heating that utilises solar power, rather than electricity from the mains. While they can have a significant upfront installation cost, solar air conditioners are much cheaper to own and operate in the long term, as they utilise virtually unlimited energy provided by our sun.

Types of Solar Air Conditioning

Three main types of solar air conditioning are:

Solar Thermal Hybrid System

This solar air conditioning system saves power by using solar energy to assist in the functions of the compressor (to compress and heat up fluid).

The system makes use of a solar collector panel such as those used in solar water heating systems to heat the fluid. However, this system requires electricity to run. In addition to that, there are questions regarding its usability at night.

Solar PV Hybrid System

This type of solar air conditioning works in much the same way as the solar thermal hybrid system. However, this system transforms the sun's energy into electricity through photovoltaic technology.

The electricity is then used to charge the batteries (when the sun is shining). When it is dark, the system switches to backup battery power. At the same time, the system uses alternating current (grid electricity) to charge its batteries.

Evaporative Solar Powered Conditioners

When it comes to solar air conditioning, no other solar air conditioner beats the environment friendliness of solar powered absorption air conditioners.

This system does not have a compressor (which is the most electricity consuming part of the air conditioner. Instead, it uses a heat exchanger, a condenser that uses water as a refrigerant instead of CFC and absorbents such as lithium bromide or ammonia.

This system only uses electricity to power the fans, chillers and hot water pumps meaning that it has significantly lower operating costs than a conventional air conditioner.

However, because of the dependence of these systems on solar energy, their usability through the night is limited. In addition to that, these systems are best for commercial purposes.

Benefits of Solar Air Conditioning

Research indicates that conventional air conditioners can constitute up to 30% of a household's electricity demand. For the average American household, this is often expensive. By using solar air conditioners, however, it is possible to reduce the electricity bill and make crucial savings.

Solar air conditioners also have the benefit of being environment-friendly. Their ability to save on electricity translates to fewer carbon emissions into the environment.

In addition to that, solar air conditioning systems such as the solar powered absorption system do not use CFC as a refrigerant. Instead, they use water.

Thirdly, solar air conditioners are sustainable. These systems use the very same heat from the sun to cool people from the high temperatures caused by the sun's heat.

It is impossible to ignore the fact that solar powered air conditioners reduce ones dependability on the grid.

Over-reliance on the grid means that crucial functions such as air conditioning would be cut off during a black-out.

With solar air conditioning, however, these functions continue for as long as the sun is up.

References

- Disadvantages Of Solar Energy - Conserve Energy Future". Conserve-Energy-Future. Retrieved 2015-11-01
- Applications, photovoltaic-solar-energy: solar-energy.technology, Retrieved 29 March, 2019
- Bergesen, Joseph D.; Tähkämö, Leena; Gibon, Thomas; Suh, Sangwon (2016). "Potential Long-Term Global Environmental Implications of Efficient Light-Source Technologies". Journal of Industrial Ecology. 20 (2): 263. Doi:10.1111/jiec.12342
- Foster, Peter (2009-10-18). "$10 solar-powered lamp to help the poor". Telegraph.co.uk. Retrieved 2015-12-16

- Mcdermott, James E. Horne; Maura (2001). The next green revolution : essential steps to a healthy, sustainable agriculture. New York [u.a.]: Food Products Press. P. 226. ISBN 1560228865
- Solar Powered Flashlight". Solarpoweredflashlight.org. Archived from the original on 2012-03-11. Retrieved 2011-07-08
- What-is-solar-air-conditioning, appliances: canstarblue.com.au, Retrieved 30 April, 2019

PERMISSIONS

We would like to thank the editorial team for lending their expertise to make the book truly unique. They have played a crucial role in the development of this book. Without their invaluable contributions this book wouldn't have been possible. They have made vital efforts to compile up to date information on the varied aspects of this subject to make this book a valuable addition to the collection of many professionals and students.

This book was conceptualized with the vision of imparting up-to-date and integrated information in this field. To ensure the same, a matchless editorial board was set up. Every individual on the board went through rigorous rounds of assessment to prove their worth. After which they invested a large part of their time researching and compiling the most relevant data for our readers.

The editorial board has been involved in producing this book since its inception. They have spent rigorous hours researching and exploring the diverse topics which have resulted in the successful publishing of this book. They have passed on their knowledge of decades through this book. To expedite this challenging task, the publisher supported the team at every step. A small team of assistant editors was also appointed to further simplify the editing procedure and attain best results for the readers.

Apart from the editorial board, the designing team has also invested a significant amount of their time in understanding the subject and creating the most relevant covers. They scrutinized every image to scout for the most suitable representation of the subject and create an appropriate cover for the book.

The publishing team has been an ardent support to the editorial, designing and production team. Their endless efforts to recruit the best for this project, has resulted in the accomplishment of this book. They are a veteran in the field of academics and their pool of knowledge is as vast as their experience in printing. Their expertise and guidance has proved useful at every step. Their uncompromising quality standards have made this book an exceptional effort. Their encouragement from time to time has been an inspiration for everyone.

The publisher and the editorial board hope that this book will prove to be a valuable piece of knowledge for students, practitioners and scholars across the globe.

INDEX

www.ingramcontent.com/pod-product-compliance
Lightning Source LLC
LaVergne TN
LVHW080807220926
848485LV00011B/385
* 9 7 8 1 6 4 1 7 2 6 5 4 2 *